The Fire Island National Seashore

The Fire Island National Seashore

A History

Lee E. Koppelman
and
Seth Forman

State University of New York Press

Published by
State University of New York Press, Albany

Printed in the United States of America

For information, contact State University of New York Press, Albany, NY
www.sunypress.edu

Production by Judith Block and Eileen Meehan
Marketing by Susan M. Petrie

Library of Congress Cataloging-in-Publication Data

Koppelman, Lee E.
The Fire Island National Seashore : a history / Lee E. Koppelman, Seth Forman.
p. cm.
Includes bibliographical references and index.
ISBN 978-0-7914-7341-2 (hardcover : alk. paper)
ISBN 978-0-7914-7342-9 (pbk. : alk. paper)
1. Fire Island National Seashore (N.Y.)—History. I. Koppelman, Lee. II. Title.

SB482.N7F66 2008
333.91'70974725—dc22 2007025000

10 9 8 7 6 5 4 3 2 1

In Memorium

David Spirtes
April 15, 2004

I find the great thing in this world is not so much where we stand, as in what direction we are moving. To reach the Port of Heaven, we must sail sometimes with the wind and sometimes against it, but we must sail, and not drift, nor lie at anchor.

—Oliver Wendell Holmes

Contents

Illustrations

Tables

Figures

Photographs

Acknowledgments

We would like to thank Kristen J. Nyitray and F. Jason Torre of Special Collections and University Archives at Stony Brook University Libraries for their extraordinary help in providing documents relating to the late United States Senator Jacob Javits, of which they are the excellent custodians.

We would also like to thank Ms. Diane Abell, Landscape Architect/Park Planner for the Fire Island National Seashore; Paula Valentine, Director of Interpretation of the Fire Island National Seashore. In particular, we would like to thank Mr. Steve Czarniecki, the Curator of the William Floyd Estate of the Fire Island National Seashore, for his help and support in navigating us through the complicated and fascinating history of the Fire Island National Seashore. More than any other single individual, this book was made possible by his efforts.

We are also indebted to several of the citizens of Fire Island, who graciously shared their insights with us. In particular we would like to thank, Gerald Stoddard of the Fire Island Association; Murray Barbash, builder of the Dunewood community; and Irving Like, Esq., for their valuable input.

We also acknowledge the support and editorial improvements provided by Dr. Michael Rinella, Senior Acquisitions Editor and Judith Block, Senior Production Editor at SUNY Press, and Lani Blackman, freelance professional copyeditor.

We are also deeply indebted to the Administrative Assistant of the Center for Regional Policy Studies, Edie Jones, who tirelessly prepared the several drafts of the manuscript and provided editorial input in addition to the yeoman's task involved in readying the book for our friends at SUNY Press.

It is important to note that the views and conclusions contained in this document are those of the authors and should not be interpreted as representing the opinions or policies of the U.S. Government. Mention of trade names or commercial products does not constitute their endorsement by the U.S. Government.

Chapter 1

Fire Island—Historical Background

Brief Overview of Fire Island History

Fire Island has been the location for a wide variety of historical events integral to the development of the Long Island region and the nation. Much of Fire Island's history remains shrouded in mystery and fable, including the precise date at which the barrier beach island was formed and the origin of the name "Fire Island." What documentation does exist, however, tells an interesting tale of Fire Island's progression from "Shells to Hotels," a phrase coined by one author to describe the island's evolution from an Indian hotbed of wampum production to a major summer resort in the twentieth century.[1] Throughout its history Fire Island has contributed to some of the nation's most important historical episodes, including the development of the whaling industry, piracy, the slave trade, and rumrunning. More recently Fire Island, home to the Fire Island National Seashore, exemplifies the late twentieth-century's interest in preserving natural resources and making them available for public use.

The Name. It is generally believed that Fire Island received its name from the inlet that cuts through the barrier and connects the Great South Bay to the ocean. The name Fire Island Inlet is seen on maps dating from the nineteenth century before it was attributed to the barrier island. On September 15, 1789, Henry Smith of Boston sold a piece of property to several Brookhaven residents through a deed that stated the property ran from "the Head of Long Cove to Huntting East Gut or Fire Island Inlet." The alternate name given in the deed reflects the influence of the Dutch in the Long Island region prior to the eighteenth century, who at various times used the terms Inlet New Gut, Nicholls Gut, Nine-Mile Gut, Huntting East Gut, and Huntington East Gut, the word "Gut" stemming from the Dutch term *gat*, meaning "gate."Another strange Dutch name for the Fire Island Inlet was "Sictem Hackey," apparently an allusion to local Indians, called Secatogues after a segment of land in Islip.[2]

The historian Richard M. Bayles argues that the use of the word "fire" in the term Fire Island Inlet stems from a misreading of the word "five," which would have been the number of barrier beach islands that existed off the Long Island coast in 1688. Other scholars believe the name derives from a misspelling of the Dutch word *vier*, meaning four, as "fier" which was then further transmogrified into "fire." Some speculate that "*fier*" could have been a bastardization of the word "fire." Still another theory conjectures that the name arose from actual fires on the island itself. There is a legend that swamp fires smoldered on the island for many years. The beaches were also used by Indians who lit fires to signal the mainland, as well as by early whalers who often boiled down whale blubber into whale oil. It is likely that fisherman and whaling crews stoked fires along the inlets to guide boats into the bay at night before the days of lighted buoys. It is also possible that "wreckers" lit fires on the beach to lure ships ashore, where oral tradition tells us crews were murdered and robbed and the ships plundered for their cargo.[3]

The Indians called the beach itself "Seal Island" because of the large numbers of seals that once wintered there. In the 1840s, the appellation "Raccoon Beach" enjoyed a certain vogue among hunters who found the beach rife with the furry canoids. Names given the island by early cartographers include "Beach," "South Beach," "South Beach of sand and stones," "East Beach," and, in 1839, "Great South Beach," a name many consider the island's official name, at least for that portion of it that lies within the jurisdiction of the Town of Brookhaven. In 1935, the Town of Brookhaven informed the U.S. Bureau of Geographic Names that the portion of the island within Brookhaven's jurisdiction should be called "Great South Beach." But that didn't stop master builder Robert Moses from calling for "restoration and protection of Fire Island" in his 1938 "Moses Plan" and influencing modern usage in favor of that appellation. The National Geographic Society, in August 1962, made a formal application to the Board of Geographic Names, United States Department of the Interior, to investigate which name—Great South Beach or Fire Island—was proper. The Board made a study and found, in decision list no. 6301 made on February 12, 1963, that while the barrier beach was known to more people as Fire Island both names were correct. For all practical purposes, however, the naming of the Fire Island National Seashore in 1964 rendered the debate over use of the name "Great South Beach" moot.[4]

Social Evolution of Fire Island. While Fire Island remained mostly uninhabited prior to the middle of the nineteenth century, its resources were exploited by a wide variety of people for a wide variety of purposes. Indians made use of the island for manufacturing wampum and hunting whales for at least several hundred years before European settlers arrived in the 1700s. In fact, the Indian names for Long Island were "Seawanauke," which means "Isle of Shells," and "Paumanauke," which means "Isle of Tribute." Both names refer to the purple portions of hard clam shells that Long Island Indians carved out, strung on thongs, and used as both currency and tribute to the more powerful Naragansett Indians of Connecticut. Indians killed whales from the beaches on Fire Island, a skill they taught to European settlers. The settlers arrived in the middle of the seventeenth century and exploited the island primarily for four different resources: whales, oysters, fertilizer, and fowl. Fertilizer, produced largely from menhaden or slow-moving horseshoe crabs, became a major regional industry in the nineteenth century. Fishing and fish processing also made an appearance. Fleets of oceangoing vessels cast their nets off shore and returned to Fire Island laden with fish. A processing plant was established on a site just northeast of the lighthouse. The Great South Bay's south side became renowned for its commercial oyster beds in the late 1700s. By the early 1800s, the beds had been harvested so intensely that oysters were planted and cultivated before the Civil War to prevent the loss of a valuable industry. On the eastern end of Fire Island commercial hunters made a handsome living on waterfowl that were plentiful on the island. This activity was later prohibited by law. After March 23, 1688, when Chief Tobaccus sold the rights to all whales that came on to the barrier beach to the Town of Brookhaven, the white settlers improved on the whaling techniques the Indians taught them. Fire Island played a significant role in the advent of the shore whaling industry that had its epicenter in East Hampton, Southampton, and Bridgehampton.[5]

Fire Island and the Fire Island Inlet played important roles in more infamous activities as well. Before the decline of slavery on Long Island after the American Revolution, slave runners sailed through the inlet to Great River or Islip where some estates had facilities for keeping slaves. Fire Island may also have been home to stockades used to hold slaves temporarily. By 1800, there were approximately one hundred slaves in the Town of Brookhaven. After slavery was legally abolished in New York State in 1827, "blackbirders" kidnaped free blacks and took them south to be sold back into slavery.[6]

Pirates also favored the island for the temporary storage of loot. Hunters, fisherman, and other potential witnesses on the island could be spotted from afar through the treeless, flat terrain, making the island ideal for illicit activity. Ship "wreckers"—people who lured ships to land only to kill their occupants and plunder their loot—were common on Fire Island. Jeremiah Smith, perhaps the first inhabitant of Fire Island, built a house near what is now Cherry Grove that he used for wrecking activities. Women apparently also took part in wrecking. Island legend has it that "two brutal women" waylaid a ship in 1816 and killed ten of the crew.[7]

The more than four hundred ships that were reported in distress off Fire Island between 1640 and 1825 gave rise to calls for the construction of a lighthouse. In 1825, the first Fire Island lighthouse was built on land the federal government bought from the state on the edge of the Fire Island Inlet. The lighthouse cost ten thousand dollars and consisted of eighteen lamps making a complete revolution every ninety seconds. This "high tech" contraption succeeded in reducing the amount of serious wrecks, but in the summer of 1850 the bark "Elizabeth" ran aground and stirred calls for a better lighthouse. The Elizabeth wreck was a highly publicized event because among its victims was Margaret Fuller, the literary editor of the *New York Tribune* and early advocate of women's rights. In 1858, a replacement lighthouse, which still stands today, was erected, slightly to the north and east of the original. This lighthouse was larger and more effective than the original, rising 166 feet above average sea level and beaming a light flash every minute.[8]

Photo 1.1. Fire Island Lighthouse (Courtesy of Fire Island National Seashore)

Another response to the frequent mishaps at sea were the creation of unmanned rescue huts put up by volunteer groups as early as 1805. Companies that insured ships knew they could recover most of the cargo if help got to a wreck quickly. So the industry pressured Congress to set up lifesaving stations on Long Island and New Jersey, which they did beginning in 1848. The volunteer efforts were effective, resulting in roughly one hundred thousand lives saved over the decades. By the 1870s the federal government started allocating money to hire crews. Beginning in 1887, the newly created U.S. Life-Saving Service created twenty-three manned stations in Suffolk County. Fire Island hosted eleven of these rescue stations, many of which participated in heroic rescues.[9]

The steam vessel era eliminated many of the dangers faced by oceangoing craft. By the 1920s, the baymen of Fire Island turned their attention to rumrunning. The Eighteenth Amendment to the U.S. Constitution was passed in 1919 and after a grace period of one year, enforcement of prohibition began. Fire Island was ideal for bootlegging for the same reasons it was attractive to pirates: bootleggers could land liquor directly on uninhabited stretches of beach for transhipment to shore or for pickup later. Rum Row, a fleet of liquor-laden vessels floating three miles out in international waters, was easily accessible from the island. Ship captains unloaded liquor in burlap sacks over the sides of their boats and floated the bounty on life preservers to helpers waiting on the beach.

The Sunken Forest, a three-hundred-year-old holly forest protected by two sets of sand dunes at Sailors Haven, was a prime destination for rumrunners looking to stash the precious liquid their associates had floated to shore. South Shore icehouses used by local fisherman were easily converted to liquor transport stations. To grease the wheels, bottles were handed out to local residents, fishermen, and enforcement officials alike. Sometimes deliveries were made directly to Fire Island communities. Prohibition was credited with reviving the small community of Water Island, which had almost disappeared during World War I. Other Prohibition Era features of Fire Island included the Bailey Stage, opposite Center Moriches, and the Saltaire Casino, which served liquor openly.[10]

Prohibition was a blip in the trajectory of Fire Island's development much as it was for the rest of the nation. Fire Island had previously begun to discard its dubious reputation in favor of one as a world-class summer resort. The primary catalyst for this transformation was David Sturgis Sprague Sammis, the man who built the Surf Hotel. Sammis was not the first hotelier on Fire Island. That distinction belongs to Felix Dominy and his wife. Dominy was the keeper of the Fire Island Lighthouse, a craftsman, and a small innkeeper. But it wasn't until 1855, when Sammis purchased 120 acres of grazing land east of the Fire Island Lighthouse for erecting his impressive hotel, that Fire Island became the toast of society. Over a span of forty years Sammis added to his original 100 foot long structure and boardwalks until the walks actually extended for several miles and the hotel buildings had a frontage of more than six hundred feet. The hotel at its peak could accommodate fifteen hundred guests and stood out from other resorts of the time for having gas lamps throughout. By the 1870s, historian Richard M. Bayles could remark that the Surf Hotel was "holding rank among the celebrated watering places of the Atlantic seaboard."[11] The hotel became famous for its masquerade balls and Sammis sported a fleet of "sail and catboats" for the enjoyment of his guests. Sammis facilitated the transport of his guests to the hotel by building a trolley line from Babylon to the dock where his steam yacht carried them to Fire Island. This allowed many of New York's most prominent men to leave their families at the hotel, to work in the city during the day and return to the hotel in the evening, perhaps a two-hour trip in all.[12]

Fire Island Land Uses. There are today two principal types of land use on Fire Island that are, in the main, oriented toward summer recreation: residential developments, and federal, state, county, and municipal parks. Each of these land uses provides for the common needs of different user-groups and for the different needs of individual user-groups.

Photo 1.2. The Club House, circa 1870 (Courtesy of Long Island State Park Commission)

The Development of Communities. The demise of the Surf Hotel during the cholera epidemic of 1892 set the stage for the development of private communities. Until that time, land on Fire Island had been owned by a small number of people. In 1693, a loophole in the Town of Brookhaven's first patent allowed William Tangier Smith to claim forty square miles of Long Island, including all of Fire Island, on which he established his estate named the "Manor of St. George." In gratitude to those who made his claim possible, he relinquished 175 acres on the west side of the Mastic River to the Poosepatuck Indians, who still occupy a somewhat smaller piece of land there. After a series of actions by the state legislature, jurisdiction over a westerly portion of Fire Island was given to the Town of Islip. In December 1834, a commission of representatives from the towns of Brookhaven and Islip drew jurisdictional boundaries that correspond roughly to what exists today. While some of Smith's heirs held onto the property left to them (mostly on the eastern half of Fire Island), one heir, Henry Smith, decided in 1779 to sell his property to "twenty yeomen of Brookhaven" for two hundred pounds. Until David Sammis began buying proprietary shares of Fire Island in 1845, no one bothered with the details of land ownership there. Conflict ensued when Sammis acted on his belief that he had enough shares to establish a hotel on 120 acres east of the lighthouse. A lawsuit—*Green vs. Sammis*—was finally initiated in 1871 challenging Sammis' title. In 1878, after years of intensive searches for shareowners, court appointed arbitrators divided the beach into seventy-eight parcels, the first of which Sammis was allowed to keep for his hotel, and the second which Sammis shared with Benjamin Sire. Sammis' Surf Hotel was eventually bought by New York State in 1892 to serve as a quarantine and screening center for ships emanating from Europe during the cholera epidemic there. Today, the state still owns the land as part of Robert Moses State Park, the first of New York's State parks.[13]

The development of private communities was made possible by the Great Partition of 1878, and the development of these communities have their origins in a wide variety of circumstances. A few communities emerged out of the housing needs of laborers. Others grew out of real estate promotion. Some expanded gradually around hotels, and one grew out of worker housing required by a fish factory. Fire Island's community developments consist mostly of summer residences, commercial establishments, and related utilities. They are situated primarily at the western end of the island, within the boundaries of the Fire Island National Seashore. Collectively, these communities provide the bulk of the island's recreational facilities (mostly private), virtually all of its overnight accommodations, and most of its commercial establishments. Fire Island is very much a study in extremes. The bright, contemporary

Photo 1.3. Davis Park T–Dock, 1945 (Courtesy of Titus Kana, www.FireIsland Vision.com)

designs so publicly displayed in the mostly gay enclave of Fire Island Pines lies only a short distance from the more sedate shingled, Victorian-style cottages of Point O'Woods, the island's most restrictive community. In between the extremes there is great variety: the sturdy modern designs of Saltaire, the low-keyed mix of traditional and modern architecture of Water Island, and the understated community of Fair Harbor and its close neighbor Lonelyville. In general, the architecture of Fire Island communities is not highly distinguished, mostly due to the difficulties associated with small and irregular lot sizes, restrictive zoning and building codes, and transportation to the island of building materials. The residential areas are political subdivisions of Long Island's towns of Islip and Brookhaven and the villages of Ocean Beach and Saltaire, which provide limited municipal services.[14]

Point O'Woods, the oldest community on Fire Island, sprung from an educational and religious assembly. Lot 21 of the divided land, today home to the communities of Ocean Beach, Seaview, Ocean Bay Park, Point O'Woods, and Oakleyville, was originally bought by the Point O'Woods Association from the Chautauqua Association in 1898. The Chautauqua Association was part of an educational and cultural self-improvement movement that began in upstate New York in the 1870s and thrived for several decades. The Fire Island Chautauqua Association, started by a group of businessmen, was established on 150 acres at "'Point of the Woods' to furnish people with the attractions of a summer resort and to give them a program of variety and excellence in musical, literary, artistic, scientific, and religious subjects." After four years of operation, the association decided to sell its lots.[15]

Photo 1.4. Davis Park Trustee's Walk, 1945 (Courtesy of Titus Kana, www.FireIslandVision.com)

Point O'Woods, founded in 1894 as a religious enclave, has thrived since that time, despite having lost ten homes in a storm in 1962, and enjoys a mix of modern and Victorian architecture. Today the community maintains a level of exclusivity by fencing it's border to Ocean Bay Park (the "Fence") and maintains a family-oriented flair. A private ferry services this community which sports grand, shingled, beach houses. Point O'Woods over the years has managed to maintain the old world charm of a private and cozy beach community. It is a predominantly Protestant Club that is determinedly family-oriented, and it prohibits the sale of alcohol, a legacy of its origins in Chautauqua.[16]

Photo 1.5. Fire Island Lightship, 1932 (Courtesy of Titus Kana, www.FireIslandVision.com)

Oakleyville, lying on the bay side just east of Point O'Woods is second to Point O'Woods in age and developed out of the desire of John Oakley to do carpentry work for Point O'Woods residents. Oakley built the first house there in 1897. The Oakley's lost possession of the land but left their name to the tiny community of ten houses, no stores, no ferry, no water supply except individual wells, and no services.[17]

Lonelyville is a small community sandwiched between Robbins Rest and Dunewood that began in 1905 when the South Shore Realty Company began selling lots there. George S. King, a south shore physician and renowned chronicler of shipwrecks on the island was the first buyer. In the 1880s, Captain Selah Clock built a four-hundred foot pier out into the ocean and there established his Fire Island Fishing Company. The pier was demolished in the hurricane of 1938. Before the 1950s there were only twenty-three houses in Lonelyville, but developers built bay-to-ocean walks, and east-west crossing, and a community water system after World War II, which permitted more development on the eastern part. By 1979, Lonelyville had five hundred feet of beachfront and forty-two homes.[18]

The next location to be developed was Ocean Beach, the largest community with the most year-round residents. In 1921, the Stay-a-While Beach Estates started by the heirs of Wilmot M. Smith merged with several lots originally purchased by John A. Wilbur to form the incorporated village of Ocean Beach. In New York State incorporated villages are classified as general purpose municipalities with an elected mayor and board of trustees. Villages can also establish police, fire, water, sewage, and maintenance departments. Ocean Beach was home to Fire Island's first elementary school, which opened with ten pupils in 1918. A second elementary school constructed to educate students on the western end of the island, opened in Kismet in 1933. High school students living on Fire Island today cross the bay to Bay Shore High School. Ocean Beach is the community that was loved by old-time stars like Fanny Brice, Billy Rose, George Gershwin, John Garfield, I. F. Stone, Carl Reiner, and Mel Brooks; where Marilyn Monroe visited Lee Strasberg and said, "What a lovely place this is—it's got water all around it"; where Walcott Gibbs, the *New Yorker*'s drama critic, edited the newspaper and wrote a collection of short stories, *A Season in the Sun*, later adapted for Broadway.[19]

Wintering on Fire Island is difficult. About two hundred fifty Ocean Beach residents choose to do so, along with around one hundred seventy-five people on Kismet, fifty in Fair Harbor and another seventy-five scattered among the other communities. Ocean Beach can furnish most of its inhabitants needs, but the Robert Moses Causeway on the western end and the bridge at Smith Point in the east provide much needed points of contact for Islanders who would otherwise suffer isolation or undue risk in crossing the icy bay to the mainland. Today Ocean Beach is one of two incorporated villages on Fire Island and is very much the hub of the barrier island. It is, in fact, the de facto capital of Fire Island, if only because most vacationers to the island eventually end up there. All other Fire Island communities depend on Ocean Beach's wide variety of shops, restaurants, bars, hotels, and services, though some spurn the community's restrictions on eating or dressing at the beach, drinking in the streets, leaving glass in garbage cans, and skateboarding. According to the community's web site, "OB" tries to run a "tight ship" because it has so many visitors and summer residents. There are six hundred homes in Ocean Beach, an elected mayor, two churches, a police force, a public green, and a court that is in session every Saturday. Because Ocean Beach was mostly spared by the 1938 hurricane, many of the old and picturesque houses remain. Not far from the commercial district by the bay you will find quiet and quaint walks in this seaside resort.[20]

Ocean Beach has spawned four tiny "satellite" communities: Atlantique, Robbins Rest, Fire Island Summer Club, and Corneille Estates. Corneille Estates, the location of the island-wide elementary school, is the largest of these communities with fifty-three houses, though it is only two blocks wide. Atlantique is the smallest satellite with only sixteen houses, but it is host to a hostel of the Appalachian Mountain Club. All of these communities use Ocean Beach for shopping, services, and ferry. Like Point O'Woods, the Fire Island Summer Club owns its land and rents it to householders on ninety-nine year leases. Robbins Rest derived its name from the Robbins family of Bay Shore, who started the forty-one home retreat in 1925.[21]

The other incorporated village on Fire Island, Saltaire, takes its name from an English town on the River Aire named for Lord Salt and was started by the purchase of land by the Fire Island Beach Development Corporation in the fall of 1910. Aggressive marketing resulted in the fast growth of Saltaire, the home of Clam Island, Fire Island's only natural protected harbor. By 1911, over a hundred homes were built and by 1912 enough boat owners were members that they launched a boat owners association and a yacht club. Saltaire officially incorporated in 1917, making it the first incorporated village on Fire Island. Originally, the Fire Island Beach Development Corporation demolished Saltaire's sand dunes in an attempt to sell property based on ocean views. That resulted in the devastation wreaked on the community in the 1938 hurricane, which destroyed half the community and killed four people. Saltaire received funding from the Works Progress Administration and the New York State Teacher's Retirement Fund for repair. The village recovered in the postwar decades to such a degree that the zoning code was changed in 1965 to limit development to 470 houses.

Because of the destruction to the community visited by storms, this middle-class family community is characterized by housing distinguished by the dichotomy of "post–1938" and "pre–1938" construction patterns. Saltaire was a major player in the development of the Fire Island Association in the 1960s and two of its residents, George Biderman and Norma Ervin, have been presidents. Houses are generally larger here than elsewhere on Fire Island and there are some wonderful examples of older cedar shingled beach houses on the bay side. Saltaire has two churches. There are no restaurants or bars.[22]

The establishment of a restaurant on Fire Island in 1868 by Archer and Elizabeth Perkinson would eventually lead to the development of Cherry Grove, the "first, and for years the only gay-controlled geography in the United States." The Perkinsons purchased a piece of property after the partition of 1878 upon which they built a hotel. The property sported a small group of cherry trees, from which the community derived its name. Like the developers of Saltaire, the Perkinsons leveled the sand dunes so that guests could see and walk to the ocean more easily. As early as the 1920s, gays and heterosexuals shared the resort. But the hurricane of 1938 devastated the community. The devastation of the storm reduced the appeal of the hotel to heterosexuals but continued to attract homosexuals because it afforded them the social protection of the island's remote physical location. After 1938, the hotel was turned over to Ed Duffy, who renamed it Duffy's Hotel. After World War II, Duffy's attracted large numbers of visitors. By the 1950s, Cherry Grove represented " 'the fulfillment of the American dream for gay people,' a place to escape the city's heat and noise and socialize without the fear of violence or need of subterfuge." As the summer capital of homosexual America, Cherry Grove continues to shape the public image of what it means to be gay: white, affluent, socially exclusive, young, promiscuous, artistic. The Duffy Hotel was destroyed by fire in 1956, but the sand dunes were restored and the community has not lost a home for more than twenty years.[23]

The forty house community of Water Island also had its origins in a late nineteenth-century hotel. Edward Ryder is widely credited with having built the "White House" hotel in 1890 and rumor has it that Theodore Roosevelt spent weekends there with his family. By 1906, the Caldwell Realty Company started to develop the area, but growth slowed during World War I. Prohibition again made the White House a hot spot for visitors. The hotel was rehabilitated by William Hauck, who served liquor openly and permitted limited amounts of gambling. Predictably, the repeal of prohibition and the Great Depression resulted in the foreclosure of Hauck's property. Today, residents of Water Island fiercely resist the accoutrements of modern life. The community has no stores, no restaurants, no public water supply or services of any kind. Only grudgingly did residents accept electricity in the 1970s, and up until recently the community was reachable only by private boat or beach taxis from Davis Park. The latter form of transportation has been outlawed by the Fire Island National Seashore and ferry service is now available.[24]

The community of Seaview grew out of Gil Smith's fish factory. The community originally housed workers for Smith's factory beginning in 1895, but after Smith closed the factory he began to sell lots to vacationers who made it clear there would be no more fish processing plants in the area. Seaview property deeds were restrictive. Along with prohibitions against the construction of a fish factory, many deeds in Seaview banned the sale of property to Jews, a restriction lifted in 1928. In the 1960s, the community experienced a construction boom resulting in large houses of modern design. Seaview is now a family resort community with 342 houses, a general store, a liquor store, and the first synagogue established on Fire Island. It has no facilities for day-trippers.[25]

Fair Harbor was born of property bought in the 1920s on partition lot six by Captain Selah Clock and George Weeks. It was originally meant to be a community for working people, and it attracted mostly people from Bay Shore who built modest housing for themselves. The community was devastated in the 1938 hurricane but made a comeback in the 1950s and 1960s and today there are more than three hundred houses, two grocery stores, a liquor store, a post office, and Le Dock, an elegant restaurant. The community discourages day-trippers, but there are renters. Most of the community, however, is made up of families, many of whom provide the support for a yacht club dock and a Swimming Association.[26]

Photo 1.6. First Airmail to Fire Island, 1929 (Courtesy of Titus Kana, www.FireIslandVision.com)

In the 1920s, the three thousand feet separating Seaview and Point O'Woods was turned into Ocean Bay Park, with a forty-house development erected by developers for working-class summer vacationers. The community had to be rebuilt completely after the Great Depression and the Hurricane of 1938, a job that was left to the Flynn family. This family of a father and five sons established a casino on the bay and a hotel built out of an old Point O' Woods coast guard station and an army barracks floated over from the mainland on a barge. Development accelerated, attracting day-trippers who used Ocean Bay Park's snack bars, restaurants, hotels, motels, and other commercial activities. Today, Ocean Bay Park is a very diverse community with many different groups of interest. There is a large contingent of group houses, the Fire Island Hotel and Resort, and nightlife is strong at Flynn's and other places right on the bay. Ocean Bay Park is filled with college and recently graduated young people. "Party hardy" is the general theme. There are also many longtime residents that care deeply about their community. Ocean Bay Park is most widely known for it's Happy Hour, which starts at about 4:00 P.M. on Saturdays and goes on into the early morning hours, something about which families in the community have expressed reservations. The community boasts perhaps the best fire department on Fire Island, a result of closely spaced, easily flammable wooden houses. The community has a grocery store, three bar/restaurants, two pizza places, and a bagel shop.[27]

Kismet is the most westerly private community on Fire Island, beginning only four hundred feet east of Robert Moses State Park. The community began in 1925 when the Kismet Park Corporation purchased almost all of partition lot three from the Sammis estate. As early as 1933, Kismet had enough year-round residents to open a school for children serving the western part of Fire Island. While the hurricane of 1938 swept away most of the permanent dwellings, there are today 250 homeowners, mostly residents of Islip or Bay Shore who rent them out. The three developments within Kismet attract mostly singles who rent houses as a group. The families that do occupy homes seem to live in harmony with the single "groupers." Being so close to the Robert Moses Causeway makes wintering in Kismet easier than in other communities of Fire Island, but only for those who obtained licenses to drive vehicles on the island prior to 1976. Since that year, the Town of Islip has issued no new licenses.[28]

Two postwar communities grew up in tandem beginning in 1945, when four developers floated a large building over from the mainland and set up a restaurant/bar called "Leja Beach Casino" next to Davis Park, a Brookhaven town park and marina. Home development began shortly after and continued in the 1960s, when the eastern part of the

Photo 1.7. Leja Beach Ferry, 1945 (Courtesy of Titus Kana, www.FireIsland Vision.com)

community broke off from the western part, calling itself Ocean Ridge. Today, the area has 250 houses. The marina itself now has two hundred boats, and has given rise to a third segment of the population called "live aboards." Live aboards stay on their boats for two weeks, enjoying the community's amenities, patronizing the casino, and shopping at the sole grocery store. Davis Park has long been popular with singles and home to the "sixish," a popular innovation in which a single residence would be randomly chosen as the site of a "bring-your-own-bottle" party. The sixish went out of style soon after parties moved outdoors "leaving a trail of collapsed decks as testimony to its success." The presence of the Catholic Church, which holds an annual ball and a Labor Day art show, a medical facility called "Bedside Manor," and the Davis Park Association would seem to lend enough stability to the community to make it a mature family summer colony. But the public facilities and park supplied by the Town of Brookhaven make it attractive to day-trippers and probably ensures the continuation of Davis Park and Ocean Ridge as a youth favorite.[29]

The community of Dunewood is distinct for being the only Fire Island community planned and built by a single contractor, Maurice Barbash and Irwin Chess. Barbash and Chess launched this community of one hundred small homes in 1958 on twenty acres of land. Barbash was an environmentalist who devoted a good portion of his life to the preservation of Fire Island. Because the homes were mostly uniform in design and featured common postwar architecture, Dunewood was derided as "Levittown-by-the-ocean." Today, Dunewood is a family-oriented community consisting of about one hundred houses. There are no shops or bars and food shopping is done in neighboring Fair Harbor.

Photo 1.8. Davis Park Marina, 1952 (Courtesy of Titus Kana, www.FireIsland Vision.com)

The community considers itself family-oriented and rentals are prohibited. There is a yacht club and all activities are limited to residents.[30]

Fire Island Pines is the most flamboyant and extravagant of the Fire Island communities. Home to many thirty- and forty-year old high-paid professionals, the community boasts six hundred houses, making it the close rival of Ocean Beach in size and population. The Pines was bought in 1925 by the Home Guardian Company from Antoinette Sammis, but was not developed until 1947. While the Pines is unique for being the only Fire Island community to have a one-hundred-unit cooperative apartment complex and a large share of Fire Island's limited number of swimming pools, its distinguishing characteristic is its culture, which derives from the high proportion of designers, models, photographers, and others in the world of high fashion that summer or live there. Gay male couples make up the majority of summer residents. Fire Island Pines parties tend to be lavish affairs that begin after 12:00 A.M. and run into the late morning hours. The Pines is also home to a community center housing a medical facility, an auditorium, a library, and a post office. The community center was converted from one of the two original Fire Island Coast Guard Stations that stand in its original location, the other one being Flynn's Hotel. The Fire Island Pines Property Owners Association (FIPPOA) was formed in 1953, and was perhaps the main force behind the community's development. In 1993, the FIPPOA celebrated its fortieth anniversary with a party featuring a fifteen piece orchestra and an appearance by New York City Mayor David Dinkins. With 570 members out of a summer population of five thousand, the FIPPOA produces highly professional music, art shows and other community activities. The Pines Conservation Society has as its mission to build up the dunes and beautify the Pines community. The Pines is also home to a nineteenth-century building that was once part of the Coast Guard station known as Lone Hill.[31]

The Eastern Half. The communities of Fire Island developed on the west end of the barrier island and occupy only 20 percent of the twenty thousand acre barrier beach. In the eastern portion the heirs of

Photo 1.9. Summer homes on Fire Island, 2003 (Courtesy of Suffolk County Planning Department)

William Tangier Smith retained title to large sections until ownership passed to the Fire Island National Seashore in 1964. No established communities ever materialized on the eastern end of the island and when the U.S. government acted to nationalize the beach it was already mainly in public hands. Suffolk County and the Town of Brookhaven came into a sizable portion of the land through a series of land transfers beginning in the early 1700s made by the William Smith family for reasons that remain murky. Other portions of the eastern end of Fire Island were inhabited by squatters, as ownership became harder to prove as time wore on. The beach on the east end was mostly marshland and not attractive to builders. Shoals at the eastern end of the Great South Bay and those in Moriches Bay made access to the island difficult. Commercial bird hunters used the land in the fall until forbidden by law. The island's hay was particularly valued.

Charles Howell leased land at Smith Point to operate a private beach between 1901 and 1910. By that time most of the older Smith property—all of the land east of Whalehouse Point to the Southampton border, with the exception of Bailey Stage—had passed into the hands of Charles F. Lynch, of Paterson, New Jersey and into Brookhaven Town custody. In 1954, Miss Eugenie Annie Tangier Smith bequeathed the remaining Manor of St. George to the Town of Brookhaven for use as a museum, along with a twenty-seven acre public park.[32] Figure 1.1 depicts the location of Fire Island communities.

Lynch had big plans for the development of what he called the "Tangier Club," but the plans were aborted when the construction of a large bridge was necessary to carry them out. Only a small wooden draw bridge linked the island to the mainland, and that remained usable only by pedestrians. The bridge was destroyed by a storm in 1917 and went through many reconfigurations. In 1931, Suffolk County approved a Robert Moses plan to spend $31 million on various projects, including a substantial bridge at Smith Point. Work did not begin immediately and plans were shelved until after World War II. Finally, in July 1959, the bridge and a public bathing beach on land acquired by the county

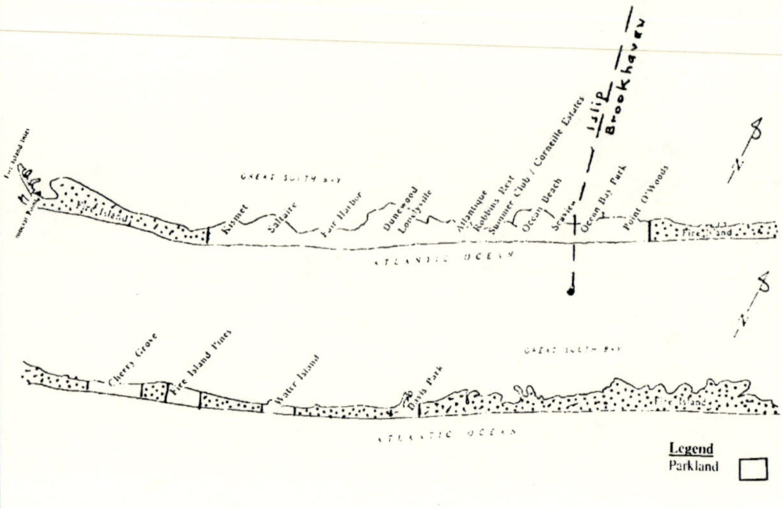

Source: Center for Regional Policy Studies, Stony Brook University, Stony Brook, N.Y.

Figure 1.1. Fire Island Communities and Parklands

were opened. The county soon after bought land from Walter Shirley and extended its popular beach and attendant parking lots.

Today, the County of Suffolk owns all the land east to the Moriches Inlet except for Great Gun Beach, a small public facility operated by the Town of Brookhaven, and a 16.4 acre lot at Whalehouse Point called "Ho Hum Beach" which the village of Bellport purchased in 1963. When the U.S. Government created the Fire Island National Seashore in 1964, it bought the property owned by Nancy Ljunqvist and her sister. With this act, the last pieces of the twenty-four miles of beach property once owned by William Smith passed into public ownership.[33]

For a summary of the housing units on Fire Island see Table 1.1.

Table 1.1. Fire Island Housing Units—2000

Community	Total (1962) Housing Units	Total Housing Units	Occupied Housing Units	Seasonal Housing Units	Other Vacation Units
Saltaire	128	401	14	342	45
Ocean Beach	605	595	61	523	11
Fire Island (Outside Village)	1,727	3,157	138	2,976	43
Fire Island Total	2,460	4,153	213	3,841	99

Source: U.S. Bureau of the Census 2000, *Fire Island National Seashore: A Proposal* (Washington, DC: U.S. Department of Interior, 1964).

Chapter 2

Fire Island Erosion

Shifting Sands and Shoreline of a Barrier Island

Introduction

It is not precisely clear how barrier beaches such as Fire Island form. The 282 barrier island complex of the United States eastern seaboard has been shaped primarily by marine deposition.[1] At the height of the last Ice Age (a.k.a. Wisconsin Glaciation) sea level was some 130 meters lower than today and the shoreline was many miles south of its present position. The glacial ice did not get as far south as the present position of Fire Island, but sand washed out of the melting glacier and blanketed the area (where Fire Island now exists) with sand ("outwash sands"). As the glacier melted sea level rose, rapidly at first, probably too rapidly to form barrier islands from the outwash sand. However, about six thousand years ago the rate of sea level rise slowed down, which probably allowed the formation of a barrier island miles to the south of Fire Island's present position. There is some evidence offshore that a barrier island existed there about thirty-five hundred years ago. It is a matter of scientific debate whether this formation slowly migrated northward with rising sea level (called barrier island "roll-over") or was periodically downed in place and reformed further north until it reached its present position. Although there's shore erosion the position of Fire Island seems to have been stable, at least over the period for which historical records exist (170 to 200 years).[2]

Fire Island is continually remolded and shifted by waves, wind, and currents. Wave action is the process which contributes most to the transformation of Fire Island. During winter months waves undercut the dunes and distribute the dune sediment upon the berm. More than eight thousand waves pound the Fire Island beaches every day, with a force reaching two thousand pounds per square foot. This relentless pounding makes the barriers extremely unstable, subject to drastic alteration as a result of storm events and longshore transport. Longshore transport of beach sediment is an important process, for it controls the amount of sediment which is available to the beach. The dominant direction of littoral transport of Fire Island beach sediment is westward. Estimates of the rate of littoral transport differ widely, but one widely used estimate gives a rate of 450,000 cubic yards annually at Fire Island Inlet.[3]

Inlet formation and sediment transport processes drive barrier beach migration and are integral to the stability of Fire Island. Inlets function as large sinks of sand in the near shore system. Generally, Long Island's south shore barrier beach survives by sand moving east to west. The westward elongation of Democrat Point at Fire Island Inlet provides a striking manifestation of the dynamic character of the barrier.

The position and number of south shore barrier beach tidal inlets have changed frequently in part due to catastrophic storms that have cut new inlets. Some of these inlets have filled naturally due to the rapid movement of large volumes of littoral drift from the east to west along the shore; others have been maintained through channel dredging and jetty construction. Man-made structures, such as jetties and groins, interfere with the longshore transport of beach sediment. In effect, jetties and groins cause the up-current beaches to accrete, while down-current beaches erode. With few exceptions, the most acute beach erosion problems are the result of the interruption of sand transport patterns at inlets. Jetties and groins, as well as numerous other man-made structures, can contribute to these erosion problems if they are not supplemented by sand bypass and beach replenishment techniques.[4]

It is believed that both the Moriches Inlet in the east and the Fire Island Inlet in the west, which form the outer boundaries of Fire Island, broke through sometime before 1690. Before these inlets appeared, the offshore areas now covered with water were meadows, saltwater marsh, and swamps. Early maps indicate that Fire Island was once a peninsula attached to the mainland near present-day Quogue. Moriches Inlet first appears on charts from 1620, although from 1836 forward no inlet is charted in the region until three storms in March 1931 broke through the barrier again. The Fire Island Inlet appears on Robert Ryder's map of 1670. Since 1690, it is estimated that Fire Island Inlet has moved roughly ten miles to the west of its original location near present-day Point O'Woods. This estimate is derived from observing that the Inlet now flows about five miles west of the Fire Island Lighthouse, which stood at the edge of the Inlet when it was first constructed in 1825.[5]

The 1834 shoreline mapping by the United States Coast and Geodetic Survey is the first reasonably accurate indication of the overall high-water shoreline position for Fire Island. Continued westward migration is shown by the high-water shorelines of 1867, 1909, 1924, and 1939. A landward shift in direction (retrogradation) is seen in the comparison of the 1867 and 1909 shorelines. A small retrogradation and almost the same amount of lengthening occurred between 1909 and 1924. One geomorphologist estimates the western tip of Fire Island grew westward at an average rate of 212 feet annually between 1825 and 1939. The geophysical evidence shows that the barrier island system has migrated continuously, albeit intermittently, for about the past ten thousand years from an offshore position of about one hundred feet below present mean sea level.[6]

Studies in long term trends in shoreline position have found that the Fire Island barrier is really two dynamically different areas, separated by an older transition zone. The western half of Fire Island is locationally stable, vertically building, but narrowing. From the Watch Hill area eastwards, the barriers are moving sporadically but rapidly landward as storm-induced inlets provide back barrier sediments for marsh growth and subsequent upbuilding by overwash with minimal wind contribution. As New York Sea Grant coastal processes and facilities specialist Jay Tanski has explained, "historical maps and aerial photos show that the patterns of shoreline change are highly variable in this area. Some areas appear to be stable or even gaining sand over the last century while others are eroding at tens of feet per year." Mr. Tanski and his colleague Henry Bokuniewicz note the consensus of the science is that "western Fire Island has been axially stable for hundreds of years," but that eastern Fire Island faces more formidable erosion difficulties. A detailed study from 1970 found the annual landward erosion rates of the Fire Island dune system to be between three and five feet per year between 1959 and 1979, and that a maximum in excess of sixty-five feet of landward erosion occurred from July to December in certain parts of the island.

The same survey showed that approximately 56,800 cubic yards of material was eroded from the dune system and that approximately 18,000 cubic yards of material was eroded from the berm along a 1,660 foot section of beach. Average shoreline positions vary on the island about 183 feet annually.[7]

Causes of Erosion

All changes occurring along the shoreline are the result of man, wind, and tides. Some of the more common short term beach changes are those brought about by summer and winter wave conditions. Long low waves of summer are able to move sand onshore, while the steeper storm waves of winter move sand offshore. Wave action and longshore transport of sediment is an important process on Fire Island, since this process controls the amount of sediments which is available to the beach. Wind does not directly influence erosion, and most dunes are stabilized by vegetation. No significant alteration of the berm, dunes, or vegetation, which stabilized the dunes, has occurred as a result of human activity. But a stable dune system can also be a threat to back bay habitat and species.[8]

Storms are another short term erosion activity and the greatest immediate concern for developed communities on Fire Island and on the south shore of Suffolk County. Storm waves can damage beaches and are also responsible for opening new inlets across the barrier beach. Between 1635 and 2002, 237 storms of varying severity hit Fire Island. Before 1900, no one lived on the island and so storm impact made little difference. After the island's residential development in 1920, though, storms that rearranged the beach impacted many lives. It was a series of storms in March 1931 that cut through the Moriches Inlet again and made Fire Island a "real" island, something it had not been since the Shinnecock Inlet had closed up fifty years earlier.[9]

Photo 2.1. Storm damage at Westhampton, 1938 (Courtesy Long Island State Park Commission)

Erosion Control—A Sorry Saga

The defining event in the creation of the Fire Island National Seashore was the disastrous hurricane of September 21, 1938. Ernest S. Clowes vividly recounted the impact and horror of the storm which resulted in heavy loss of property and human lives.

> Soon after three o'clock the situation on the beaches became critical, especially on that long strip from Shinnecock Bay to Moriches Inlet where the dunes were mostly low and had at their backs a succession of bays and canals. And, as the storm approached from a bit west of south and the trend of the coast eastwards is a little north of east, the center reached Westhampton before it did points farther east, [*sic*]. By three the sea there was all over the beaches and beating and breaking at the foot of the dunes. By half-past three it was breaking over and through the dunes at many places and sometime toward four o'clock the final catastrophe occurred. Before the onslaught of that terrible tide, itself perhaps ten to fifteen feet above the normal height and crested with breakers towering fifteen feet higher or more, the whole barrier of the dunes crumbled and went down save for here and there where a higher dune or a strong bulkhead held. In a few minutes along the stretch of beach from Quogue Village to Moriches Inlet there remained of 179 summer homes only 26 battered shells of houses of which hardly a dozen will ever be habitable again.[10]

Shortly before 4:00 o'clock the dunes gave way before the terrible force of the roaring surf, houses collapsed, cars were tumbled like leaves, some of the stauncher houses were floated intact and whirled crazily in the core of the hurricane. Geography changed as new inlets were pushed through by the angry sea demanding an outlet for its force. For over two hours there was no difference between the Atlantic Ocean at its worst and the usually placid Moriches Bay, as the latter was swollen by the inrush of lashing water.[11]

This legendary storm generated 120 mile per hour winds and caused damages totaling $6 million. Out of 1,219 existing homes on Fire Island, 265 were lost and an additional thirty-two were damaged. The greatest havoc took place east of Fire Island, resulting in loss of life as well. Hurricanes and northeast storms are not rare events in the history of Long Island. National Weather Service data since 1886 indicate that the Island has been directly impacted by hurricanes and tropical storms. Northeast storms, causing significant water-related damage,

Photo 2.2. Storm damage to Superintendent's quarters at Fire Island State Park, 1938 (Courtesy of Long Island State Park Commission)

Photo 2.3. Point O'Woods, September 22, 1938 (Courtesy of Long Island State Park Commission)

occur nearly every year. Unusually severe storms occur in the area about three times every century. The 1938 hurricane was followed by a rash of storms in 1944, 1950, 1953, 1954, 1960, 1962, 1968, 1969, 1985, 1991, 1992, 1993, 1996, and 1999, with the 1962 storm bringing the most destruction: 100 houses lost, 130 houses damaged, and total losses amounting to $16.5 million. The Point O' Woods community was destroyed. More recently, the storm of 1992 reshaped the shore and overwhelmed natural sand accretion. Along with flattened dunes, a threatened breach on Fire Island, damage at Gilgo Beach, large-scale erosion at Robert Moses State Park, and a serious breach threat near Shinnecock Inlet in Hampton Bays, that crisis caused a second tidal inlet to develop known as Little Pike's Inlet.

Photo 2.4. 1962 Storm Damage. (Courtesy of Long Island State Park Commission)

Photo 2.5. 1962 Storm Damage. (Courtesy of Long Island State Park Commission)

Photo 2.6. Storm Damage in the Flood Plain, 1962 (Courtesy of Long Island State Park Commission)

This was the first time since 1931 that Westhampton Beach and Fire Island have been bordered by more than the single shallow, narrow Moriches Inlet opening. The United States Corps of Engineers (USCE), with state and county support, closed Little Pike's Peak in 1993 by installing 1,800 linear feet of buried steel sheet pile and pumping 1.5 million cubic yards of sand onto the breach.[13]

Natural replenishment of the beach, achieved by relinquishing certain dune areas to the waves, only to rebuild them in certain other places, is complicated by the presence of humans. Humans impact the beach primarily through two activities: modifying dune fields by building homes, parking lots, and roads on them; stabilizing inlets through dredging and the construction of jetties, groins, seawalls, and bulkheads. Both activities contribute to or are impacted by erosion. The modification of dune fields by man-made structures prevents the transport of sand into the back barrier and tidal marsh, and puts lives and property at risk. Inlets are stabilized for the purposes of navigation and tidal flushing. Nevertheless, the structures and dredging that are required to maintain the inlets contribute to erosion by blocking the natural littoral drift of sand, depriving down-current beaches of replenishment. These two human activities, in turn, give rise to the two most pressing coastal needs: protecting against tidal flood damage and erosion caused by hurricanes and storms; and halting the erosion of beaches and dunes on the island due largely to the maintenance of inlets.

Coastal areas on Long Island have experienced dramatic residential and commercial development and change in recent years. As a result, the Island's south shore is far more vulnerable to storm-related damage and potential loss of life today than it was sixty-six years ago when the devastating hurricane of September 21, 1938 destroyed Westhampton Beach and other shoreline communities. This increased vulnerability exists despite the fact that early warning systems and hurricane forecasting techniques are now in place. The Flood Disaster Protection Act of 1973 (P.L. 93–234) requires that every community identified by the U.S. Department of Housing and Urban Development as having areas of special flood hazard participate in the National Flood Insurance Program. Fire Island has been identified as having special flood hazards and approximately 80 percent of the island's residences are covered by this federally subsidized insurance.

These storms, abetted by misguided governmental responses, have exacerbated acute and chronic erosion problems occurring on Fire Island. From 1938 to the present, all past management efforts can be characterized as short-term responses to crisis situations at specific locations. Immediately after the 1938 hurricane, the USCE initiated the first study to protect the barrier beaches on the south shore of Long Island from storm and storm-related erosion havoc. Robert Moses added his recommendation that an ocean highway be constructed atop a twenty foot high dune from the Fire Island Inlet easterly to the Suffolk County beach property at Smith Point. No action was taken. Several reasons can account for this. First, the number of homes on the barrier beach was limited and the property owners did not have any political clout. Second, the inland population of Suffolk County had no interest in their tax monies being spent to protect beach property owners' investments—particularly in view of the fact that many Fire Islanders were from the New York Metropolitan area, and not Suffolk residents. The third factor was the very limited nature of Suffolk County government.

Until 1960, the county was considered a rural county controlled by a Board of Supervisors made up of the Supervisors of the ten towns, each with an equal vote. Thus, the supervisor from Shelter Island representing less than two thousand population exercised an equal voting power with the supervisors from western towns such as Babylon, Islip, Huntington, or Smithtown, whose populations were many times larger. Since the prime responsibility of the supervisors were to their towns, the attention paid to their duties as county legislators was minimal. Most governmental services were the responsibility of the towns, including welfare, police, planning and zoning, assessment, and the majority of highway construction projects and road maintenance. When coupled to the conservative antitax posture of the general population, no constituency existed to ameliorate the threat to Fire Island property owners.

This is of little moment in terms of public policy were it not for the inherent danger to human life and property values. Prior to the construction of homes and the concern of owners for stabilizing what in essence should be a movable island, the migration of the beach was left to natural causes. After all, the barrier system survived for millennia without interference or assistance from human attempts to lock the land in place and are counter to the action of natural forces. Left free of human interference, the barrier was constantly reformed, albeit in a different position.

From a conservation viewpoint, the spit could survive without interference. However, houses cannot. Each devastating storm washes houses out to sea or causes their collapse. The islanders understood this and seemed willing to assume these risks. They also demanded that governmental action be taken to minimize these hazards. In response, the USCE has proposed an extensive dredging and fill program to rebuild the dunes. This immediately raised the objections of conservationists and shellfish interests who maintain that irreparable harm would result ecologically and financially. This area of dispute continues to the present.

Erosion Control Efforts

Government officials have long recognized the instability of the Fire Island barrier beach and have instituted intermittent programs to mitigate it. The cutting or burning of grasses, the grazing of cattle, or the cutting of timber on Fire Island was outlawed as long ago as 1784 by the New York State Legislature in an attempt to stabilize the sand dunes. Over the years shoreline damage caused by wave and tidal action has resulted in the construction of shore protection devices and beach filling. These kinds of projects have been financed by private individuals, beach associations, local municipalities, the State of New York and the Federal government. These projects have often been constructed on a piecemeal basis without a comprehensive evaluation of their potential effects on large segments of shore.[14]

A stone jetty was built by the federal government south and west from the 1939 shoreline of Democrat Point at the Fire Island Inlet. The jetty brought an abrupt halt to the lengthening of Fire Island. Instead, beach sand was trapped by the jetty, causing a seaward movement (progradation) of the high-water shoreline as the new beach filled in. Completed in 1941, the jetty reached its impounding capacity in 1948; then littoral drift sand again began to spill over into Fire Island Inlet in large quantities. Northward shift of Fire Island Inlet, because of this natural filling, caused tidal currents to scour sand from Oak Beach on the north side of the inlet.[15] Between 1946 and 1959, more than one million yards of sand were pumped onto the Oak Beach shore in an attempt to restore those beaches.[16]

In April 1958, the USCE announced their initiation of a study to rehabilitate and maintain the south shore coast from the ravages of storm induced shoreline erosion. Their plan was publically released on October 26, 1959.[17] It called for a $36.9 million initial cost to relocate thirty-five million cubic yards of sand to raise the height of the dunes, increase the depth of the beach, and to vegetate the dunes in order to help stabilize them. Their plan also recommended the construction of as many as fifty groins to be placed between the Fire Island Inlet and Montauk Point. The construction cost would be shared by the federal government assuming 51 percent, with the remaining balance of 49 percent to be paid by New York State and Suffolk County.

It was also estimated that the annual cost of maintenance would amount to $336,000, or approximately 1 percent of the construction cost.[18] This latter expense generated immediate negative reactions from Suffolk County Board of Supervisor members who generally were not in favor of the entire project; they particularly objected to the County being saddled with a continuous maintenance cost. This lack of support after years of study and public hearings extended the status quo of ignoring erosion problems during the two decades since the 1938 hurricane.

However, even if enthusiastic support from residents and elected officials existed, progress toward a workable solution would possibly take years just to achieve approval. Part of the inertia can be attributed to the difficulties inherent in getting all levels of government—federal, state, county and local—to reach a consensus on methodology, overall cost, cost sharing, and future financial responsibility for maintenance. The difficulties are further compounded at the federal level itself. A brief description of the administrative hurdles that must be cleared helps to illuminate the bureaucratic morass that must be successfully waded through before any project reaches solid ground.

Once the design engineers complete their technical work, which itself can take several years, their plan is then submitted to the Chief of Engineers, who in addition to reviewing the work may also amend it. This would then require some revision by the design staff. Assuming technical agreement within the USCE is reached, the proposal is then distributed to a variety of state and federal agencies who exercise direct or collateral interests and/or responsibilities that could be affected by the project. If any opposition arises, it could kill the project.

Assuming this hurdle is surmounted and the commentary received by the USCE is supportive or at least neutral, the project is then submitted to Congress. There is no time limit imposed on the congressional response. The first action that must be taken is by the Public Works Committee of the U.S. House of Representatives that must clear the project and include it in the Rivers and Harbors Omnibus Bill for authorization. The next action requires passage of the bill including the project by Congress. Once passed, the bill is then referred to the President for his signature. If all of these steps are positively affirmed, action is still not assured.

One major step still has to be achieved. Congress has to reconsider the project and pass funding legislation in an appropriations bill. Authorization without appropriation cannot go anywhere. Each step in the process outlined in these previous paragraphs can involve a time lapse of months, if not years.

A response from the residents of Fire Island to the frustration of seeking relief from the constant erosion that threatened their homes was to seek palliatives that might delay if not reverse the inexorable loss of sand. This took form in a campaign to collect and deposit Christmas trees.[19] Although this somewhat feeble effort could not remediate the erosion forces created by periodic storm events and the continual effect of littoral drift that moves sand in a westerly direction, it did serve to keep a public focus on the problem.

On the Oak Beach segment of Fire Island, including the developed communities of Kismet, the Village of Saltaire, Fair Harbor, Lonelyville, Atlantique, Robbins Rest, Corneille Estates, Village of Ocean Beach, Seaview, Ocean Bay Park, and Point O'Woods, the most common beach stabilization measure employed is the use of snow fencing to trap windblown sand in an attempt to enhance or create dunes. The Village of Ocean Beach has constructed two small groins and has created dunes using unsightly concrete rubble as core material. Some communities have sought permits to scrape sand off the summer berm and use it to create dunes.[20]

Central Fire Island, including the communities of Cherry Grove, Fire Island Pines, Water Island, Blue Point Beach, and Davis Park, and the parts of the Fire Island National Seashore that includes the Sunken Forest nature sanctuary and Sailor's Haven, a public marina and beach facility, also rely primarily on snow fencing to trap windblown sand in an attempt to enhance or create dunes. The community of Fire Island Pines has installed plastic seaweed in an offshore location in an attempt to trap sand to build an offshore bar to dissipate wave action. Over one-half of this segment of Fire Island remains undeveloped, predominantly shrubland vegetation, fresh and tidal wetlands in the back bay areas, as well as a double dune system in some locations (e.g., at the Sunken Forest).[21]

It would appear logical, therefore, to expect that the 1959 program suggested by the USCE would have received immediate broad support. That this did not occur can be explained by examining the conflicting positions held by mainland Suffolk County residents, appointed and elected town and county officials, and Robert Moses and the Long Island Park Commission.

A strong element of public mistrust of Moses' recommendations and Corps responses goes beyond the decades old efforts of Moses to create a dune highway almost the length of Fire Island—albeit the road issue alone would suffice to arouse homeowner antagonism. The history of the construction of the Fire Island Inlet channel, although not of effect on Fire Island, does typify the citizens skepticism over claims made by government officials.

Fire Island Inlet. Moses continually claimed that his solution to erosion of the barrier beaches would be the stabilization of a sufficiently high continuous dune line by placing a roadway atop the dune. The Jones Beach State Park and Ocean Highway was cited as proof of his concept. What Moses didn't share with the public was the full range of activities carried out by the Long Island State Park Commission that were necessary to achieve beach protection. This included an almost constant nightly sand replenishment to Jones Beach, after the public had left for the day. Thus, the road didn't prevent erosion or hold the dune in place.

Sand for the Jones Beach replenishment was drawn from the channel on the bay side of Jones Island. An additional potential source of sand was in the choked Fire Island Inlet, where serious erosion was a threat to Cedar and Oak Beaches. The Long Island State Park Commission successfully convinced the USCE to initiate the dredging of a Fire Island Inlet channel in the mid-1950s.[22] The issue of stabilizing the Fire Island Inlet was first debated in 1949. The new channel, to initially cost $2.3 million, would create a channel one thousand feet wide to a depth of sixteen feet, generating three million cubic yards of spoil (sand). This material would be pumped onto Cedar and Oak Beaches. In addition, some of the sand would be used to replenish Jones Beach. The USCE also recommended an erosion control program of beach replenishment from Fire Island to the Hamptons for an additional $26 million. The boating interests at Captree and the residents and users of Cedar and Oak Beaches were strongly in support of the Inlet projects. However, there was little support for the larger projects. Opponents, including Suffolk County officials, claimed that merely building a perched beach without the construction of properly spaced groins, was a futile waste of money. Experience had amply proven that the sand would be lost at the first storm event. In fact, Suffolk County engineering experts claimed that an effective project with sufficient groins would cost $62 million, not the $26 million cited by the USCE.[23]

It wasn't until June 5, 1959 that the project for the Inlet was actually begun. The first funding of $841,000 was appropriated in 1958, with an expected completion date of October 30, 1960.[24] On December 5, 1959, a major segment was completed of a large earthen dike that sealed off the original Oak Beach channels where the erosion to the Oak Beach had lost forty-four homes in the period from 1956 through 1959 as the result of storms. Moses proposed that a permanent dredge at the southwest end of Fire Island at Democrat Point be established in order to permanently preserve the channel from filling in.

However, the Inlet project sold to the public as an improvement to boating was controversial from the outset. Congressman Stuyvesant Wainwright had continually attacked Moses since the project was first seriously proposed in the mid-1950s. After the Inlet channel and dike were completed, he more strenuously challenged the wisdom of the Long Island State Park Commission, claiming that the new inlet was a navigational hazard.[25] A new organization of boaters was formed specifically to oppose the channel placement.[26] The Babylon Tuna Club's Fire Island Inlet Safety Committee expressed the view that the 2,200 foot long dike actually caused the water in the inlet to flow more rapidly, thereby increasing the menace to boaters.

These challenges were substantiated by the Coast Guard, who warned mariners to only use the channel when the tides were slack. While this recommendation would obviously enhance safety, it would, if followed, curtail boating activity through the channel. In response to these charges, that the channel was a navigational hazard, the Long Island State Park Commission and the USCE rallied to justify and support their actions. Charles W. Boyce, who was an engineer at the Commission, defended the channel in concert with Chief of the USCE, General Emerson C. Ipschner, who rebutted Wainwright's charges. He based his support on information he received from his Chief Engineer for the Long Island region, General Thomas Libscomb.[27] Their rebuttals were also supported by other boating organizations, who were in favor of the new channel, which was considered an improvement to the originally silted-in inlet. These groups included the Captree Boatmen's Committee for the Stablization of Fire Island Inlet, the Captree Fishermen's Council, and the Fire Island Inlet Committee.

An additional element was added to the debate by another newly formed advocacy group, the Fire Island Waterways Association. Their chairman, John Suydam, wanted a new channel to be cut through Fire Island, directly from the ocean to the bay just west of the lighthouse, which he claimed would provide greater safety than the new channel through the Fire Island Inlet.[28] This proposal may have arisen from the historical existence of the Fire Island Inlet at this location a century earlier. Their suggestion was immediately countered by the USCE, who said their solution was worse and would be more hazardous than the newly completed channel. The main point was that the ocean tidal change averaged 4½ feet. The change in the Great South Bay was on the order of six inches. Any direct channel through Fire Island would change this dynamic and possibly result in south shore mainland flooding.[29]

Despite protestations to the contrary from the USCE and the Long Island State Park Commission, there was a recognition that the recently completed inlet channel needed correction. The Commission's response was to claim that the new channel was only a temporary solution.[30] They claimed that they always intended the final route of the channel would hug the western tip of Fire Island and would lie in a north-south direction instead of the east-west direction of the existing channel. They also claimed that their original plan called for the removal of seven million cubic yards of sand, not the 3.7 million cubic yards that was mentioned earlier. Of course, the additional work would cost several million dollars more. The USCE added their mea culpa, stating their intention to move the channel marker buoys one-half mile north to an area of calmer water. It appears that the Park Commission's excuse was revisionist in that there is no evidence that they informed the public at any time of a long-range, multiple-stage implementation plan. In fact, critics observed that the proposed ultimate channel could have been dredged in the first instance. Figure 2.1 depicts the constructed (temporary) channel, the new entrance suggested by the USCE, and the ultimate entrance.

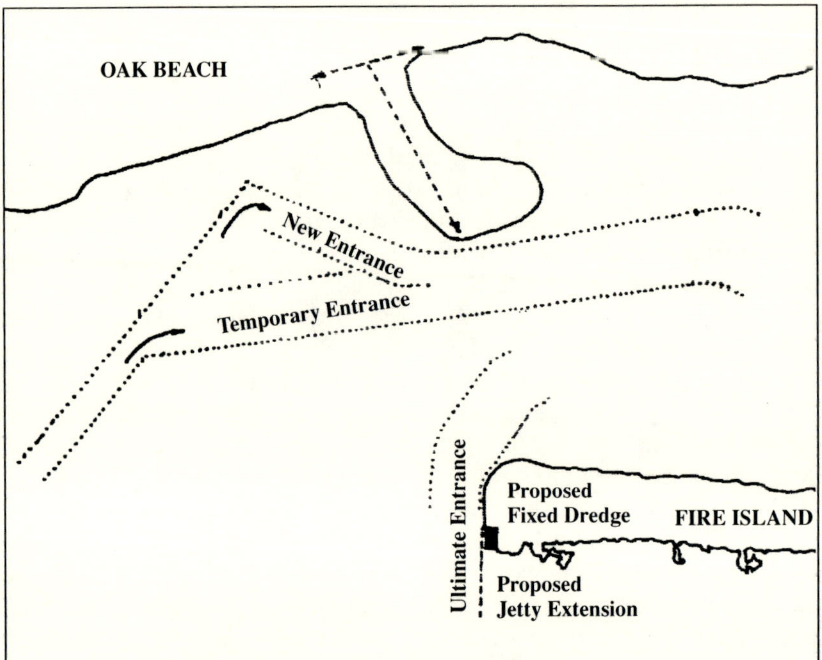

Source: Center for Regional Policy Studies, Stony Brook University, Stony, Brook, NY.
Note: Photo map shows plans for Fire Island Inlet Channel

Figure 2.1. Fire Island Inlet

USCE Reformulation Plan

During the first week in March 1962, a major northeaster's storm resulted in significant damage on Fire Island, especially at Point O'Woods. Ironically, on the first day of the storm, *Newsday* reported that the USCE Reformulation Plan, which was released in 1959 and which had received congressional authorization for the expenditure of $38.2 million in July 1960, still had not received an appropriation measure for the federal government's share of 51 percent of the cost. The USCE indicated that action for the appropriation would be predicated on the commitment from the State of New York and Suffolk County to guarantee their financial participation of 24 percent each.[31] The USCE estimated that if everything worked well, and there were no delays in the process, the project could take more than a decade to complete. Three years earlier County Coordinator Cermak of Suffolk County indicated the County's opposition to the project, claiming the requirement for the County to pay $9 million was too excessive.[32]

Photo 2.7. Storm Damage from 1962 Noreaster (Courtesy of *Newsday*, March 1962)

Photo 2.8. Watch Hill Area, 1962 (Courtesy of *Newsday*, March 1962)

The only support for the project came from Fire Islanders. Gilbert M. Serber, chairman of the Fire Island Erosion Committee, called for agreement on the project despite the demand of county officials that the federal government assume more than 51 percent of the cost.[33]

At the least, one positive environmental accomplishment was achieved. Private citizens, who had formed the Sunken Forest Preserve, Inc., succeeded, after almost ten years, in buying the seventy-three acre Sunken Forest between Point O'Woods and Cherry Grove. This endeavor was started by Ernest G. Osborne in 1950. James N. Dunlop, who served as president of the organization, purchased the thirty-six acre forest and an equal acreage of buffer land.[34] The dedication was held on August 13, 1960. Yet, mitigation of erosion on Fire Island continued to prove elusive.

Optimism began to surface in August 1963 when the USCE announced that the long awaited erosion control project was expected to commence in two months. After a hiatus of twenty-two years, the first phase of the forty-two mile, $38 million project was ready for implementation. Agreement was reached between Suffolk County and the USCE to erect a twenty-one groin installation between the fourteen mile stretch from Moriches to Shinnecock Inlets.[35] The cost for this segment was estimated at $23.9 million. This meant that the original cost estimate of $38.2 million had now grown to $60 million.

The optimism was short-lived. The Board of Supervisors, on August 5, 1963, postponed action because of the increased costs. Supervisor Evan Griffing from Suffolk's smallest town, Shelter Island, served as the Chairman of the Board. He opposed the project on the grounds of cost and what he felt was an improper expenditure of public money to protect private property.[36]

Several factors, when viewed in the total context of the governmental participants, make it apparent that what appeared to be a flip-flop on the part of the Board of Supervisors, was not the case. These factors included a change of Suffolk County government, from one in which the part-time supervisors exercised legislative and executive powers (as they did in their roles at the town level), in addition to the judiciary; to one with an elected County Executive. The League of Women Voters took up the cause of County government reform, and they produced a 125-page report which recommended that Suffolk become a charter form of government, with an elected county executive and a professionally staffed planning department under a planning commission. The Democratic party remembered the controversy created by H. Lee Dennison's highway reform study. The study found the Republican controlled Suffolk County Department of Highways spent over $200 million without constructing a single mile of properly engineered roadways. The report resulted in Dennison's discharge, but several years later the Democrats offered him the nomination to run for the office of supervisor of his hometown, Brookhaven. He said he would run, but only for the position of Suffolk County Executive. His campaign was based on reform; professionalism of county personnel instead of political patronage, public policy based on planning criteria rather than partisan political decisions, and greater involvement of the citizenry in the affairs of county government. This new charter form of county government began January 1, 1960, with the election of H. Lee Dennison. In addition to the added separation of powers, the Board itself changed from one of Republican domination to one of Democratic control. This overthrow resulted from a series of land scandals involving improper sale of tax lien properties. In the late 1950s, Governor Averill Harriman appointed a Special Prosecutor, who brought about a number of indictments. One Suffolk County judge went to jail; the consultant to the County Planning Board committed suicide. The newspapers ran daily expose stories and the voting public, who were overwhelmingly registered as Republicans, determined to "clean house." H. Lee Dennison, who was a registered Republican, was a professional engineer with a strong commitment to planning. Most of his career, from his coming to Long Island in 1925 until the 1950s, was with the Suffolk County Highway Department. His political naivety cost him his job and since his report was considered an attack on his agency and the entire County administration, he was also blacklisted, making it difficult for him to earn a living.

One of the numerous targets Dennison addressed was the control of erosion. As a result of his close friendship with Robert Moses, who saw in Dennison a kindred spirit and strong ally, and one who supported Moses' road proposals as part of erosion control. Dennison, at the beginning of his administration, also contacted the USCE to finally get action on the USCE 1959 Plan. He had a good relationship with the USCE, having served as a Major in the USCE during World War II. Since the Board of Supervisors were now controlled by the Democrats, he had no difficulty getting their support for virtually all of his proposals.

However, the Democratic control of the Board of Supervisors lasted for only one term. The election of November 1961 returned control to the Republicans. With the change, Dennison's support evaporated; the majority of the Board were just as opposed to the expenditure of County funds in 1962 as they were in 1959.

Nevertheless, erosion control was the second prong of the dual objectives of Fire Islanders. First, to derail Robert Moses' plan for a dune highway, and second, to get federal support to finance an effective erosion control program.

The first objective was achieved in 1963 when Governor Nelson Rockefeller, who was at odds with Moses, declared that the State had no money to build the road.

The First Steps. The first phase of the 1959 Reformulation Plan was limited mainly to Westhampton beach, east of the Moriches Inlet. In concept, the Corp's plan called for the widening of beaches along the developed areas between Kismet and Mecox Bay to a minimum width of one hundred feet at an elevation of fourteen feet above mean sea level. It also included raising dunes to an elevation of twenty feet above mean sea level from Fire Island Inlet to Hither Hills State Park, and at Montauk and opposite Lake Montauk Harbor by artificial placement of suitable sand. The project authorized the construction of up to fifty groins and periodic beach nourishment subject to a determination of need.

An initial placement of eleven groins was constructed in 1965. Four additional groins were built in 1969 and 1970. Two additional groins were built at Georgica Pond in East Hampton. No work was included for Fire Island. The New York District of the USCE also prepared plans for the construction of six more groins to complete the original plan for twenty-one groins extending to the Moriches Inlet. However, New York State withdrew its financial share requirements until 1973 due to a moratorium on capital expenditures. When they wanted resumption in 1973, Suffolk County now had a new County Executive, John V. N. Klein, who was opposed to groins.

The project has been the subject of criticism from its initiation. Environmentalists were opposed to the placement of groins as being counter to leaving the barrier beach subject to natural forces. Within Suffolk County government itself the Planning Director was in disagreement with the Commissioner of Public Works (DPW) over the issue of the construction schedule. Commissioner Hermon Bishop of DPW insisted that the project start at the east end of Westhampton Beach and proceed westerly. The Planning Director argued that since the littoral movement of sand was from east to west, the groins should first be built at the Moriches Inlet with a sand bypass mechanism and then proceed easterly in order to prevent scouring on Fire Island. Commissioner Bishop argued that it really didn't matter since the overall project would be built so quickly that the debate was academic.

Unfortunately circumstances proved that the Commissioner's decision was ill-conceived. After the first eleven groins were in place, there was already evidence that down beach scouring erosion was occurring. This was inescapable since the lengthy groins were interrupting the flow of sand. Two additional factors contributed to further exacerbate the problem. Earth Day and the rise of environmental activism had its genesis at the time of the completion of the first eleven groins. More public opposition was raised against "engineering the environment," which now brought more focus on the project. The second happening was a change in the County government itself. After several years of prolonged debate and court actions, the Suffolk County Board of Supervisors was replaced with a County Legislature. The former Supervisor from the Town of Smithtown, John V. N. Klein, became the presiding officer. He had strong sympathy for the environmentalist's position on groins. This resulted in a slowdown in groin construction. When he was elected County Executive in November 1971, it became clear that Dennison's engineering approach to erosion would not be continued in the Klein administration.

The consequence, resulting from this policy shift, was that the remainder of the original twenty-one groin project was aborted. This led to severe scouring along the Dune Road segment of the Westhampton Beach. The barrier beach was breached. Dune Road became impassible and the Town of Southampton prohibited access to the Dune Road residents. Litigation was instituted by the affected homeowners against the County of Suffolk, claiming that the partial construction of the groins, coupled to the lack of its completion, was responsible for the threat to their properties. The County Planning Director recommended that the County condemn all the houses, remove them, and convert the entire Dune Road strip to the Moriches Inlet as a County park.

This recommendation was not accepted by the Suffolk County Legislature. A main factor in their opposition was the reality that condemnation costs would be borne by County taxpayers. Conversely, even if the litigation went against the County and remediation would have to be made, the County's financial exposure would only be 12.5 percent, with the federal and state governments paying the rest. This dragged out for years, with a final resolution taking place in the 1980s. During this entire period, Fire Islanders vigorously tried to secure support for sand replenishment, arguing that the Westhampton Beach debacle was responsible for the erosion threats on Fire Island. That argument remains active to the present time.

The USCE did attempt to rectify the problem of erosion and they filed an Environmental Impact Statement (EIS) in January 1978 with the United States Environmental Protection Agency (EPA) relative to completion of the original project. This filing was now mandatory following the passage of the National Environmental Protection Act (NEPA) of 1969, which created EPA. Their EIS was referred to the Council on Environmental Quality as unacceptable. Most of the objections of the Council concerned the proposed use of heavy structural devices such as groins to sustain inlets and the perception that not enough attention was given to the systemwide impact of such devices. The Council also expressed concern that the plan would spur greater development of the barrier beach. In particular the Council stated: "More serious consideration of the systemwide impact of groins on beach erosion and of the advisability of removing existing groins is warranted, as well as the definition of circumstances or limitations for using long-term, heavy structural devices." In June 1978, the Council recommended project reformulation to the Chief of Engineers, who in turn directed the USCE's New York District to reformulate the project. Reformulation was initiated in 1980. But difficulties between the federal, state, and county in proceeding with parts of the project in Westhampton caused the suspension of reformulation. Based on letters of intent to support an interim plan for Westhampton, the reformulation study was resumed in the 1990s.[37]

Current Efforts. Studies for a comprehensive reformulation project were initiated in 1994 in accordance with an Initial Project Study Plan dated June 1993. The goal of the reformulation study is to identify and evaluate alternative methods of providing the authorized beach erosion control and hurricane protection for the barrier island. The project study plan was revised in June 1996. The cooperating agencies, including the State of New York and the United States Department of Interior, have discussed further revisions to the reformulation study plan resulting in the current project Management Plan dated June 2000, which increases the study scope, time and cost of the prior plan. A design agreement has been signed to share the increased costs with the State of New York. As of 2004, an environmental impact statement is being prepared for the reformulation effort.

The USCE, the State of New York, and the Department of Interior have been engaged in extensive coordination to develop mutually acceptable interim plans along the Fire Island barrier island to provide protection until the implementation of the recommendations of the overall reformulation study. There are four plans for interim projects: (1) Moriches Inlet to Shinnecock Inlet (Westhampton); (2) Fire Island Inlet to Moriches Inlet (Fire Island); (3) the West of Shinnecock Inlet Interim Project; and (4) the Breach Contingency Plan. Each of the interim projects, which generally consist of sand fill only, is subject to modification, continuation, or abandonment based upon the outcome of the reformulation study. As specified in a partnering agreement between the USCE and the Department of Interior, any proposed interim project would consist of initial beach fill placement with one renourishment for a period not to exceed six years.[38]

The Breach Contingency Plan for the eighty-three mile project area was approved in February 1996, and the USCE is authorized to execute breach closures under the plan after coastal storms. The Westhampton Interim Project, which provides for a protective beach berm and dune, the tapering of existing groins (numbers 14 and 15), and construction of an intermediate groin, was completed in 1997, with beach renourishment continuing for thirty years. The West of Shinnecock Inlet Interim Plan, which consists of initial beachfill placement with two renourishments for a period not to exceed six years, was begun in the fall of 2003.[39]

The interim project proposed for Fire Island recommends fill to be placed in areas along those parts of the barrier island which do not meet the design cross section, and which are susceptible to breaching and overwash during relatively frequent events. Fill placement is included in Robert Moses State Park, the developed areas of Fire Island, and at Smith Point Park. After the 1992 storm, Governor Mario Cuomo established the Coastal Erosion Task Force to review storm damage and to work with the USCE in making recommendations. Consistent with the interim project, the Task Force recommended prompt closure of breaches, modification and filling of the Westhampton groin field, and implementation of sand bypassing at all south shore inlets. The Draft Decision Document and Draft Environmental Impact Statement were submitted for public review in December 1999 and a public meeting was held in January 2000.

Large-scale opposition to the Fire Island Interim Plan emerged quickly. The Department of Interior and several environmental groups objected to various aspects of the proposed sand filling. The Nature Conservancy, on one hand, favors a policy of keeping the barrier island completely undeveloped and fears that sand filling would encourage more development. This position is based on the belief that storm induced overwash, or barrier island "rollover" or "migration" is necessary for the health of the barrier island and for various animal species and plants native thereto, especially in the back bay habitat. The Nature Conservancy has proposed buying out all properties in the dune area for $163 million in public funds.[40]

The Department of Interior also believes that houses in the dune areas cause erosion and has recommended "that an enforceable building moratorium be in place prior to any beach replenishment actions." The Department of Interior has called for removing private homes by a variety of means, such as restricting owners to a life tenancy in their structures; requiring structures to be relocated to other areas, and condemning and purchasing developed and undeveloped lots. The Department of Interior is immediately not only concerned with thirty-five in-fill lots identified in the USCE Draft Environmental Impact Statement, but also is concerned more broadly with preventing rebuilding of property that may be damaged or destroyed in future storms. Driving the Department of Interior's concerns is the belief that: permitting the development of these lots is "the tip of the iceberg," thus, opening the door for further residential development in the Dune District; the additional residences will add to the erosion and dune degradation; and the project will promote a "psychological" acceptance of still more development and consequent increased demand for repeated beach nourishment.[41]

The policy of removing all privately owned structures and land on Fire Island was once the position of New York State and the Long Island Regional Planning Board. The Regional Planning Board's 1989 report *Proposed Long Island South Shore Hazard Management Program* promoted the idea of government acquisition of private residential structures on Fire Island in the event that substantial damage as a result of coastal erosion and flooding is sustained. "When private structures located within the Coastal High Risk Zone [including Fire Island] are damaged to a level greater than fifty percent of their replacement value due to either severe storm occurrence or long-term shoreline erosion, action should be taken to prohibit re-development in those locations. . . . Should regulation and other actions . . . fail to prevent redevelopment, government should acquire the damaged structures and private property at fair market value as a last resort," the report said.[42] But in light of local political opposition Regional Planning Board director Lee E. Koppelman later conceded that the goal "is no longer possible of achievement, if it ever was."[43]

For its part, New York State's policy of coastal "retreat" was reflected in the state's 1981 Waterfront Revitalization and Coastal Resources Act (Article 34 of the state's Environmental Conservation Law Ch. 841, Laws of 1981), which created the Coastal Erosion Hazard Area. The law prohibits construction on all land southerly of a line drawn twenty-five feet landward of the primary dune. This places large numbers of previously developed building sites off-limits to construction. But in its 1994 recommendations, Governor Cuomo's Task Force on Coastal Resources backed away from this hardline position, stating that "the state should adopt a policy of strategic retreat and selective protection. This approach would foster decisions that protect the public's interest and respect private property rights, while working within the realities of natural coastal processes." The Task Force report suggested spelling out more clearly where it makes sense and where it does not make sense to build in coastal hazard areas, and recommended "soft" solutions such as beach nourishment over "hard" solutions such as groins and seawalls.[44]

New York State and the Long Island Regional Planning Board have modified their positions on sand filling and beach nourishment largely in response to the effective political organization of Fire Island homeowners spearheaded by the Fire Island Association. This organization was formed from the Fire Island Erosion Control Committee which successfully fought the construction of a service road from the Robert Moses Causeway to Ocean Beach in the 1950s. The group was incorporated as the Fire Island Voters Association in 1961, and the corporate charter was modified again to change the name to Fire Island Association in 1981. The Fire Island Association Board of Directors comprises the mayors of Ocean Beach and Saltaire and the leaders of eighteen community associations. The Board elects a president, first vice-president, treasurer, secretary, and three vice-presidents every two years. Consisting mainly of homeowners fearful of storm damage to their property, the Association very much favored the Fire Island Interim Project. It is the position of the Association that sand fill and beach nourishment solutions to erosion are necessary because human activity along the barrier island has made it so. For the Association and its supporters, building on dunes is not the cause of Fire Island's erosion problem. Rather the historic manipulation of the Long Island shoreline in terms of groin building and inlet dredging on the eastern part of the barrier islands have blocked the littoral flow of sand that would naturally replenish sands further west on the island.

While there is no definitive study implicating the role of home construction in dune erosion, there is substantial backing for the belief that up coast groins and inlets do prevent the natural littoral movement of sand. In particular, the Westhampton groin field and the lack of regular sand bypassing at Moriches and Shinnecock Inlets, along with substantial alterations in the inlet and bay system over many years, has "produced a massive sand trap" that has literally "starved" Fire Island of necessary sediment has led to the current erosion problem.[45] In light of the complicity of man-made engineering projects in creating the erosion problem—particularly those conducted by the USCE—the Association argues that man-made solutions, particularly those performed by the USCE of Engineers, are essential to their mitigation. The Association also believes that building restrictions imposed by the state's Coastal Erosion Hazardous Area would preclude additional building on dunes that are artificially replenished, and they question the legal authority of the Fire Island National Seashore to impose a building moratorium.[46]

The Fire Island Association has been frustrated in its efforts to advance the Fire Island Interim project. As of 2004, the Westhampton groin field has been modified, but completion of inlet bypassing at Shinnecock and Moriches Inlets and beach filling on Fire Island has not taken place. In light of the Department of Interior's opposition to the interim plan, the New York State Department of Environmental Conservation decided not to respond to the Draft Environmental Impact Statement. Blaming the lack of a nonfederal sponsor for the interim project, efforts have focused on bringing all interested parties together to develop a comprehensive plan through the reformulation study, into which the interim project will be folded.

In the absence of federal action, several communities on Fire Island have established erosion control districts, which have levied taxes for the purpose of dredging off shore sand and replenishing vulnerable dunes. The community of Fire Island Pines has spent upward of $10 million dollars since 1992 replenishing sand.[47] Walking on the dunes is now prohibited along the entire dune zone. Communities continue to plant beach grass and vast lengths of snow fences are erected to strengthen the dunes. Sometimes old Christmas trees are placed on the dunes in hopes that they will function as a makeshift snow fence. The community of Point O'Woods engages a bulldozer to push sand from the water's edge up the face of the dunes each spring.[48]

An interesting side story to this more than four decade quest for erosion control relief was the creation of a privately sponsored group to solve the problem by hiring their own engineering firm. This was a strong expression of their frustration in getting implementation from the Corps.

The New York Coastal Partnership

The New York Coastal Partnership (NYCP)[49] was founded in 1993 by a group of residents who conduct or manage local businesses. The group, acutely aware of both the beauty and the value of the coastline, recognized that winter storms demonstrated most dramatically how fragile this beauty was and how quickly this value could dissipate.

Since 1993, the issue of integrity and maintenance of the coastline received much needed public exposure and discussion. Unfortunately, the issue in their judgment has been falsely portrayed as ecology versus economics. Those wishing to maintain businesses, jobs, and homes, dependent on the viable coastal communities, appear to be pitted against others concerned with preventing overdevelopment of vulnerable shorelines. Too often, each side misses the obvious; Long Island is an ecological/economic entity. One cannot be maintained without the other.

The contention was that the state and federal government have had no realistic agenda. Their spending plans do not include investment in protecting the coastline, although there is ample federal and state authorization for such action and a demonstrated willingness of coastal communities to contribute to the cost of such protection.

The organization is based on successful efforts in New Jersey and elsewhere, which resulted in effective governmental coastal protection projects and a dedicated funding source. NYCP's goal is to explore and promote the economic viability of our coastal resources. To do this, the group set out to synthesize the available scientific and technical data, develop beach and barrier protection scenarios, and plot them against real economic activity. Dr. Timothy Kana of Coastal Science and Engineering, Inc. (CSE) was retained to do the coastal analysis.

Recommended Plan. Initial nourishment of 6,000,000 cubic yards (cy) in Year One with 50 percent placed in each of two five mile reaches (Moriches Inlet to Smith Point County Park and Fire Island Pines to Ocean Beach). Each of these beach fills will gradually feed areas to the west.

Periodic renourishment at a rate averaging 500,000 cy per year beginning in Year Six with the major portion added at the east end of Fire Island.

Construction of sand-retaining structures (e.g., groins) along a six-mile reach from Ocean Beach to Robert Moses State Park.

Dune enhancement, restoration at washovers, and breach closure on an episodic basis where dune volumes above flood levels are lacking.

Regular shoreline monitoring to document conditions, track performance of improvements, and refine future designs.

Periodic maintenance of Oak Beach shore-protection structures.

This synopsis is from the full report which is multidisciplinary, covering the specialized fields of coastal erosion, storm-tide prediction, and shoreline flooding. Chapters cover a regional sediment budget, summaries of predicted flood elevations, estimates of damage, and alternatives for mitigation and prevention of breaches. An expanded executive summary provides an overview of study methods, findings, and implications. The main report includes detailed tables, figures, and discussion of results.

CSE based the cost estimates primarily on results of the updated sediment budget and profile volume analysis as well as recent history of nourishment costs in New York. However, results from analogous projects in other states are also drawn upon to insure the approaches are feasible and the unit costs are realistic. Because an up-to-date shoreline survey is not available yet, the quantity estimates are based on older data. Experience has shown that even out-of-date profile surveys which confirm regional decadal trends (e.g., groin field accretion, jetty fillets, inlet trapping, etc.) will yield reasonable predictions of future changes. The profile volume approach to nourishment also yields quantity estimates (for underwater, beach, and dune volumes) where comparative surveys are lacking.

An important variable is the unit cost of sand. For obvious reasons, accessibility and transportation distance tend to control unit costs. To simplify the cost estimates, CSE assumed nourishment from external (nonlittoral) sources to cost at least $5 per cubic yard. This is similar to unit costs of sand delivery for the 1993 Gilgo/Robert Moses nourishment using sand from Fire Island Inlet and for the 1993–1994 breach closure at Westhampton Beach using an offshore borrow site. Small-scale projects such as dune reshaping by bulldozers using in-situ sand or lagoon dredging with harbor dredges can be performed at much lower cost, sometimes well under $2 per cubic yard.

For medium-to-large-scale projects, the consultant drew on more than twenty beach restoration projects CSE has completed since the early 1980s using a variety of borrowed sources and placement techniques. Performance reports from many of these projects, particularly those at Hunting Island, Myrtle Beach, and Deabrook Island (South Carolina), offer insight regarding loss rates as a function of project length and unit volumes.

An overriding assumption in CSE's estimates for Fire Island is that cross-shore changes will dwarf average annual changes. These cross-shore changes constitute the "beach cycle," whereby storms erode the upper beach but post-storm periods nearly restore the profile to average or "equilibrium" conditions. Long-term maintenance of eroding beaches and dunes, therefore, depends on sustained nourishment which replaces the net losses of sand over time, generally using external sources of sand. Emergency repairs, in contrast, are sometimes needed to accelerate the recovery of the beach and dunes soon after storms. Most emergency repairs can be performed cost effectively using internal sources of sand, such as beach and dune scraping by bulldozers.

South Carolina's experience after Hurricane Hugo illustrated these two fundamentally different approaches to beach restoration. In South Carolina, sixty-five miles of emergency dunes were scraped into place less than one month after Hugo. This $1.5 million project cost only $4.50 per lineal foot (1989) because it simply entailed reshaping of the profile by bulldozers. Expenditures for beach nourishment in comparison were many times greater at approximately $65 per lineal foot over a length of twenty miles. Most of the post–Hugo nourishment projects involved placement of sand from external sources in an attempt to replace the net losses attributed to Hugo. The scale of those projects was, in turn, much smaller than a ten-year requirement for nourishment at the individual sites. Emergency dunes formed the nucleus of today's ferritin along much of the impacted coast. In some reaches, however, the scraped dune quickly eroded because of a persistent sand deficit in the profile. Similar variations will occur along Fire Island, depending on whether an individual reach has a sand surplus or deficit before a storm.

For purposes of developing a more specific plan and cost estimate for maintaining the viability of Fire Island, CSE recommended two hierarchies of need. The first was based on costs and attempts to prioritize actions that, if taken early, provide the greatest return on investment (largely through reduced costs compared with later implementation). The second hierarchy is based on the impact of an action on the regional, decadal sediment budget. Clearly, the first hierarchy emphasizes low-cost alternatives whereas the second ranks large volume projects highest.

Hierarchy of Needs—Based on Costs

1. **Episodic dune rebuilding (in place) at washovers** via in-situ scraping where the nearshore profile volumes are "healthy" (e.g., Old Inlet 1994). *Cost range*: less than $10/ft; applies to twenty-five thousand feet of shoreline following the 1992–1993 storms. CSE assumes such activities will be required once per decade following major storms. *Net budget impact:* less than $250,000/10 years.

2. **Dune enhancement via scraping/sand fencing** (etc.), using in-situ sand where dune volumes are lower than desirable and the nearshore profile volumes are "healthy." Goal should be to provide 10–30 cy/ft above the 25–100 year (stillwater) flood elevation over the length of the barrier island (applicable and cost effective along certain central Fire Island reaches). *Cost range:* less than $15/ft; assume applies to approximately fifty thousand ft of shoreline. CSE assumes this activity will be required only once per twenty years. *Net budget impact*: $750,000/20 years.

3. **Breach closure** immediately after storms, using land-based equipment/nearby sand sources. Incipient flood-tidal deltas from breaches should be considered a primary borrow source. In-situ sand from the adjacent beach should be used to expedite closure. If monitoring confirms the littoral volume "borrowed" has a measurable adverse impact on adjacent areas, later nourishment from an external source can mitigate for the sand removed under emergency conditions. *Cost range:* less than $200/ft (limited reaches) would apply to a relatively small length of shoreline on the order of five thousand ft. This is assumed only once per twenty-five years after major storms. *Net budget impact:* $1 million/twenty-five years.

4. **Beach disposal of sand from navigation projects**—Beach nourishment should be the designated disposal option for most navigation projects. Poor-quality sediments (muddy or contaminated) should not be disposed on the beach. Acceptable mud content for beach disposal is less than 15 percent in most cases. Disposal should be as far as possible from inlets to reduce the tendency for sediments to recycle back to an inlet. **Additional cost** (beyond budget for the

navigation project) of booster pumps (etc.) is estimated at $2.50/cy. CSE estimates this can provide at least 33 percent of the annual maintenance nourishment requirement for Fire Island (i.e., 1 cy/ft/yr), particularly if the placement is concentrated along eastern Fire Island in connection with Moriches Inlet projects. Estimated sand volume—200,000 cy/yr. *Net budget impact:* less than $5 million/10 years.

5. **Annual maintenance nourishment for Fire Island**—CSE's estimated needs average 3 cy/ft/yr (500,000 cy/yr) of which a portion can be derived from routine navigation dredging. The remainder (approx. 2 cy/ft/yr) will have to come from other sources such as offshore deposits similar to the borrow areas used for the Westhampton breach closure and recent nourishment at Fire Island Pines, Fair Harbor, and Saltaire. Unit nourishment costs will vary principally with transportation distances. In general, maintenance nourishment should focus on hot spots and be placed as far updrift as feasible to improve longevity. *Cost range*: $5.50/cy which, when apportioned over the 165,000 foot length of Fire Island, yields a *Net budget cost:* approximately $1.8 million—which will not be required until Year Six if initial nourishment (item 7) is completed in Year One.

6. **Selectively install sand-trapping structures along western Fire Island** where there is a measurable depletion of offshore deposits and accelerated dune erosion. Groins or segmented breakwaters (the typical class of structures) should be as short as possible to allow efficient sand bypassing once the cells between structures are filled. The logical terminus of a groin field is Democrat Point; however, a detailed study of alternative groin configurations (spacing, length, crest height, and profile slope) should be conducted with the idea of minimizing the number of sand-retaining structures. Impounded sand and a trend of accretion along Democrat Point means a field of groins could be tapered and ended well updrift of Fire Island Inlet.

Analogs for this recommendation are groins placed along bulges in the shoreline of Garden City and Debidue Island (South Carolina). In these two cases, previous inlet positions produced the initial bulge because of sand trapped by ebb-tidal deltas. Inlet migration and depletion of the abandoned delta shoals caused accelerated erosion several decades later. To hold the shoreline at the bulge, several groins were built at each locality. Once each cell filled to capacity, excess sand bypassed the structures.

The purpose of groins in any setting should be to reduce the rate of erosion without increasing downdrift erosion. Therefore, care is required before implementing this solution. Nevertheless, CSE believes it is a viable alternative for western Fire Island because of the clear trend of accelerating erosion from Point O'Woods to Robert Moses State Park followed by rapid accretion at Democrat Point. A strategically placed groin field along the "bulge" would provide a secondary benefit of reduced shoaling in Fire Island Inlet.

The scale of the Garden City (South Carolina) "bulge" is much smaller than the Fire Island Inlet bulge (in the vicinity of Kismet) due to differences in the scale of the inlets. Therefore, it is expected more sand-retaining structures would be required to stabilize western Fire Island. Using Westhampton Beach's groin field (where the structures were longer than necessary) as a guide, somewhat shorter structures on the order of three hundred ft long, spaced about fifteen hundred ft apart, provide a first cut estimate of costs. CSE assumes a field of twenty such groins would be required to stabilize an approximate six-mile reach between Ocean Beach and the state park. With the installation of a field of groins, volumetric erosion rates in the protected area will decrease to a fraction of present costs. *Cost range:* \$1,000 to \$2,500 per linear foot. *Net budget impact:* \$6–15 million for twenty structures. This is assumed to be a onetime cost over fifty years.

7. **Initial nourishment for Fire Island**—CSE estimates six million cubic yards are presently required to restore a healthy dune/beach/underwater profile along 60,000 ft (~ 33 percent) of Fire Island. This nourishment is based on an estimated profile volume requirement below mean sea level (ferritin to –20 NGVD) of about 500cy/ft. This should be placed principally in the areas between Moriches Inlet and Smith Point County Park, and between Fire Island Pines and Ocean Beach. These two areas will feed downdrift reaches and improve longevity of the project. Upon receipt of updated profiles, the existing unit volumes can be compared to identify reaches with the greatest sand deficit and refine the nourishment plan. The preferred sand spirces are offshore deposits, testing at least as coarse as the intertidal beach sediments. *Cost range:* \$6.75/cy (or \$675/ft) in the nourished areas.

Hierarchy of Needs Based on Sediment Budget. There is often much debate over the impact of various shore-protection activities, but viewed from the perspective of sediment budget preservation, the following is clear: Nourishment from an external source and maintenance of littoral transport are the primary ways of favorably impacting the sediment budget along the coast. Installation of sand fences, Christmas trees, or other low-cost alternatives yield a negligible impact on the sediment budget. The following hierarchy is based on the Fire Island to Montauk sediment budget as well as CSE's experience in similar settings.

1. *Restore annual beach nourishment and dredging disposal* on beaches to 1960s levels, Particularly from Pikes Beach (Westhampton) to Smith Point County Park. This is considered more important than initial nourishment (#2) because it represents a long-term commitment.
2. Perform *initial nourishment* to achieve minimum healthy dune/beach/underwater profiles along entire Fire Island;
3. *Slow erosion rates along the "bulge"* of western Fire Island by means of sand-retaining structures;
4. *Close breaches and return flood tidal delta deposits* to the ocean side of breaches;
5. *Restore dunes in washovers*;
6. *Enhance dunes* mechanically to preserve the littoral budget seaward of the ferritin during storms (i.e., minimize washovers);
7. *Plant dune grass*;
8. *Install sand fences, Christmas trees,* and so forth.

Summary of Costs. The above-listed barrier restoration estimates have been totaled and annualized in table 2.1 for ten-year, twenty-year, and fifty-year periods (in constant dollars) for purposes of comparing with expected benefits.

Conclusion. It now appears that USCE Reformulation Plan (which has been reformulated several times since 1959) is no closer to a satisfactory solution for the Fire Island homeowners. The latest blow has come from President George W. Bush's proposed 2005 budget. The President intends to severely slash the USCE budget, which would imperil the entire Fire Island Inlet to Montauk Point Reformulation Plan. This may be welcome news to those who support nonintervention to natural occurrences, but does not resolve the concerns of Fire Islanders.

Table 2.1. Preliminary Cost Estimates over Ten-, Twenty-, and Fifty-Year Periods (current dollars)

Item	*Net Cost (thousands)*	*Frequency*	*Average Annual Cost (thousands)* 10-Year	*20-Year*	*50-Year*
Dune Rebuilding at Washovers	250	Once in 10 years	25	25	25
Dune Enhancement to Storm Tide Levels	750	Once in 25 years	75	37.5	30
Breach Closure After Major Storms	1,000	Once in 25 years	100	50	40
Beach Disposal from Navigation Projects	5,000	Over 10 years	500	500	500
Annual Maintenance Nourishment	1,800	Each year after year	900	1,350	1,620
Sand-Trapping Structures Western Fire Island	10,000	Once in 50 years	1,000	500	500
Initial Nourishment	40,000	Once in 50 years	4,000	2,000	800
Contingency/Maintenance/Monitoring	500	Every year	500	500	500
Average Annual Cost (divided by 1,000 and rounded)			7,000	5,000	3,700
Total Cost (actual-rounded)			$70,000,000	$100,000,000	$185,000,000
Average Annual Cost (actual-rounded) per foot of shoreline			$40/ft/yr	$30/ft/yr	$20/ft/yr

Chapter 3

The Creation of the Fire Island National Seashore

Introduction

Although the Congress of the United States supported the establishment of national parks as early as 1872, the creation of the National Park Service (NPS) did not occur until the administration of President Woodrow Wilson. It wasn't until 1937 that the first of the National Seashores was authorized. The next three seashores were created between 1961 and the establishment of the fifth seashore at Fire Island in 1964. The five seashores are: Cape Hatteras, North Carolina; Cape Cod, Massachusetts; Padre Island, Texas; Point Reyes, California; and Fire Island, New York. This chapter provides a brief portrayal of the evolution of the NPS from its beginning to the present, followed by a brief mention of New York State, Suffolk County and municipal parks.

The balance of the chapter traces the legislative initiatives and debates that finally led to the establishment of the Fire Island National Seashore.

Evolution of National Park Service

The concept of the national park is widely credited to the American artist George Catlin, who voiced concern about the potential harm America's westward expansion would have on Indian civilization, wildlife, and wilderness. On a trip to the Dakotas in 1832, he wrote "by some great protecting policy of government . . . in a magnificent park . . . A nation's park, containing man and beast, in all the wild and freshness of their nature's beauty!" Before the federal government would itself become the steward of any of the nation's parks, the Congress granted Yosemite Valley in California to the state of California for preservation as a state park in 1864. This was followed eight years later by the preservation of the spectacular Yellowstone country in the Wyoming and Montana territories as the nation's first public park. As there were no existing states in the region yet, Congress reserved Yellowstone "as a public park or pleasuring-ground for the benefit and enjoyment of the people" under the custody of the U.S. Department of Interior. With the support of many western railroads anxious to promote tourism, and the idealistic impulse to preserve nature at its back, Congress followed the Yellowstone model by establishing the Sequoia, Yosemite (to which California returned Yosemite Valley), Mount Rainier, Crater Lake, and Glacier national parks in the 1890s and early 1900s. The late nineteenth century also saw increasing interest in preserving prehistoric Indian ruins and artifacts on public lands, symbolized by such acquisitions as Arizona's Casa Grande Ruin in 1889 and passage of the 1906 Antiquities Act.[1]

By 1916, the U.S. Department of Interior was responsible for fourteen national parks and twenty-one monuments, but no agency or body was specifically assigned to their management. The department was also busy managing a growing feud within the conservationist movement between those who favored regulated use of natural resources and those who favored strict preservation. This conflict came to a head in 1913 when the city of San Francisco convinced the Interior Department to permit the damming of Yosemite's Hetch Hetchy Valley for a reservoir, which park supporter John Muir called "the worst disaster ever to come to any national park." At this point, the strict conservationists had no institutional representation in government. The utilitarian portion of the conservation movement had representation in the U.S. Geological Survey and the Forest and Reclamation services. After heavy lobbying by Interior Department officials Stephen T. Mather and Horace M. Albright, Congress approved and President Woodrow Wilson signed legislation creating the NPS within the Interior Department. The act made the bureau responsible for the Interior Department's national parks and monuments, Hot Springs Reservation in Arkansas (made a national park in 1921), and "such other national parks and reservations of like character as may be hereafter created by Congress." In managing these areas, the NPS was directed to "conserve the scenery and the natural and historic objects and the wild life therein and to provide for the enjoyment of the same in such manner and by such means as will leave them unimpaired for the enjoyment of future generations." Mather, who became the NPS's first director, effectively blurred the distinction between the utilitarians and the pure preservationists by emphasizing tourism in the parks and working toward increasing the number of visitors.[2]

Over the next two decades the NPS proceeded to expand its holdings and its political support, primarily through the strategy of seeking the preservation of resources in the eastern part of the United States. Up until this point, the national parks system was mainly a western phenomenon: only Acadia National Park in Maine lay east of the Mississippi. Opportunities in the east lay primarily in American history and the preservation of historic sites. In 1933, in perhaps the most significant event in the evolution of the national park system, President Franklin Delano Roosevelt signed two executive orders. Together, the orders gave the NPS control over national capitol parks and transferred national monuments held by the Forest Service (and responsibility for all monuments created thereafter) and the War Department's parks and monuments, to the NPS. In 1935, Congress confirmed the NPS's role as the leading federal agency in the field of historic preservation with the Historic Sites Act of August 21, 1935.[3]

Several other important changes took place in the 1930s. The NPS began to get involved in managing areas intended primarily for mass recreation, such as the Blue Ridge Parkway between Shenandoah and the Great Smoky Mountains national parks, and the Natchez Trace Parkway, which were designed for recreational motoring. Many of the new recreation sites the NPS took charge of were on reservoirs, which created some animosity with park purists who saw this as a joining of forces with dam builders. The NPS's other major recreational initiative during the period addressed seashores. In 1934, the NPS identified twelve significant areas deserving federal protection on the Atlantic and Gulf Coasts. Among them was Cape Hatteras, North Carolina, which Congress authorized as the first National Seashore in 1937. That Seashore, which encompasses one hundred miles of barrier islands and beaches, was acquired with the help of the Mellon family foundations and provides an outstanding natural recreational resource.[4]

During the 1930s, at the height of the Great Depression, the NPS also embarked upon the implementation of the Civilian Conservation Corps (CCC), a major New Deal program. This program employed thousands of young men in numerous conservation, rehabilitation, and construction projects in both the national and state parks. At the program's peak in 1935, the NPS oversaw 600 CCC camps, 118 of them in national parklands and 482 in state parks, staffed by some 120,000 enrollees and 6,000 professional supervisors. The CCC was instrumental in supplying the NPS with the professional civil service staff necessary to establish a permanent regional system for park administration.[5]

America's entry into World War II in 1941 brought with it severe retrenchment in park activity, as NPS director Newton B. Drury hunkered down to stave off wartime pressures to pillage national parks for timber, ore, and other raw materials. The postwar era brought new pressures on the parks, as increasing personal income, leisure time, and automobile ownership fueled a postwar travel boom for families seeking vacations and entertainment. Visits to the parks went from six million in 1942 to thirty-three million in 1950 to seventy-two million in 1960. Parks director Conrad Wirth, a landscape architect and planner, who took over operations in 1951, embarked on "Mission 66" to meet the new challenge. Mission 66 was a ten-year, one billion dollar program to upgrade facilities, staffing, and resource management throughout the system by the fiftieth anniversary of the NPS in 1966. A hallmark of Mission 66 was the visitor center, a multiple-use facility with interpretive exhibits, audiovisual programs, and other public services. Mission 66 fell short of Wirth's ambitious goals, primarily because Congress added more than fifty parks to the system during the ten-year period, stretching NPS resources in the process. Under Mission 66 the NPS resumed the National Survey of Historic Sites and Buildings, which had lapsed during the war years, to aid in planning for the system's orderly expansion. Beginning in 1960, most historic properties surveyed and found nationally significant were designated national historic landmarks by secretaries of the Interior. In 1962, the NPS launched a similar program for natural lands, resulting in the designation of national natural landmarks. The purpose of these programs was not only to mark these outstanding places as significant, but also to encourage their preservation by others, not necessarily the NPS. By 1999, some twenty-three hundred historic properties and nearly six hundred natural areas had received landmark designation.[6]

Conrad Wirth was succeeded as director by the energetic and hard-driving lawyer George B. Hartzog in January 1964. Hartzog worked closely with Interior Secretary Stewart L. Udall during the 1960s to advance an activist park policy under President Johnson's Great Society initiatives. Under Hartzog's administration, the responsibilities of the NPS grew dramatically. Ecology and "living history" programs received greater emphasis, and the National Historic Preservation Act of 1966 authorized the NPS to maintain a comprehensive National Register of Historic Places. National register properties received special consideration when federal project planning and preservation funding was doled out.[7]

But more importantly, Hartzog's 1964 management memorandum allowed the expansion of the parks system by devising three categories of park areas and assigning each area its own management standards. Prior to this memorandum recreational areas were not legally defined as part of the parks system, mostly because Congress feared recreational uses like hunting and fishing would spread and corrupt more traditional preservation areas. But in creating the "natural," "historical," and "recreational" categories, Hartzog proclaimed that each "required a separate management concept and a separate set of management principles coordinated to form one organic management plan for the entire system." Recreational areas would emphasize public use, subordinating natural and historic resource preservation. Historic areas would emphasize the maintenance of significant historical features of an area, while managing their natural values. Natural areas would emphasize the perpetuation and restoration of natural values, without neglecting significant historical features. In 1968, three separate policy manuals were published for each category of area. Two years later Congress passed the General Authorities Act, which redefined the parks system to include all areas managed "for park, monument, historic, parkway, recreational, or other purposes" by the NPS. In 1972, in response to another policy memorandum from President Nixon's Interior Secretary Walter Hickel, Hartzog further expanded the park system's domain with his National Park System Plan. The plan subdivided the three categories of park areas into more specific categories and identified which categories needed greater representation. The plan's natural history component, for example, identified more than three hundred aspects of natural history requiring initial or expanded representation in the system.[8]

During Hartzog's nine year tenure, sixty-eight of today's park units were added to the system—nearly three-quarters as many as had been added in the preceding thirty years. Overall, between 1952 and 1972 there were twelve permanent additions to the system in the natural areas category, sixty-one historical areas, and twenty-eight permanent recreational area additions. This expansion of the system was aided by several acts of Congress. The Ozark National Scenic Riverways in Missouri, authorized by Congress in 1964, foreshadowed the comprehensive Wild and Scenic Rivers Act of 1968 and the National Trails System Act of the same year, which led to the acquisition of other free flowing rivers, trails, performance facilities, urban recreation areas, and more. Additions in all categories of the parks system were aided by the Land and Water Conservation Fund Act of 1965, which earmarked revenues from visitor fees, surplus property sales, motorboat fuel taxes, and offshore oil and gas leasing for federal and state parkland acquisitions.[9]

Of great importance to natural preservation in the system during the 1960s was the Wilderness Act of September 3, 1964. For designation as wilderness an area was to be without permanent improvements or human habitation, to retain its "primeval character and influence," and generally to contain at least five thousand acres. The application of the Wilderness Act to the parks system was opposed by Wirth, who believed the NPS identified and managed wilderness areas within parks sufficiently without it. But Congress declined to exempt the parks system and by 1999 had confirmed parts of forty-five parks as wilderness.[10]

The formal adoption of the recreational category by Hartzog's 1964 plan was preceded in the 1950s when the NPS resumed shoreline studies with support from the Mellon family foundation. These studies ultimately produced documents like *A Report on the Seashore Recreation Survey of the Atlantic and Gulf Coast* (1955); *Our Fourth Shore: Great Lakes Shoreline Recreation Area Survey* (1959); and *Pacific Coast Recreation Area Survey* (1959), along with several individual studies. As a result, eight more national seashores and four national lakeshores were added to the parks system between 1952 and 1972. The NPS, in its 1955 *Atlantic and Gulf Coasts* report identified Fire Island, together with the Shinnecock Inlet area, as one of the most important remaining undeveloped shoreline areas in eastern United States. Soon after this report, a bill was introduced in the 86th Congress to establish a Fire Island National Seashore, though no action was taken. Following the severe storm of March 1962 that did tremendous damage to the shoreline area of the Atlantic Coast between New York and North Carolina, Governor Richard J. Hughes of New Jersey called for a conference of twelve state governors at Atlantic City to discuss the problem of protecting exposed shoreline areas from future storms. Secretary of Interior Stewart L. Udall urged that immediate thought be given to dedication of certain shoreline areas for public use. At his direction, a task force of state and federal officials made an aerial reconnaissance of the shoreline from eastern Long Island to Virginia. The task force report recommended that portions of Fire Island, two miles of seashore across Moriches Inlet and five miles of seashore around Shinnecock Inlet, both to the east of Fire Island, be protected for public recreation purposes. Following the report, other proposals were made to protect and develop Fire Island and several bills were introduced in the Congress to establish a Fire Island National Seashore. In response, several studies conducted by the Bureau of Outdoor Recreation and the National Park Service found that a National Seashore was warranted.[11]

Between 1973 and 1999, the period which parks historian Barry Mackintosh describes as the "rounding out" of the system, park growth exceeded the explosive growth of the preceding period. The system added 123 new parks at this time, and a few major acquisitions in Alaska in 1978 and 1980 more than doubled the system's total land area. Not that there weren't any dips and turns in the process. Ronald H. Walker and Gary Everhardt, the immediate successors to George Hartzog, attempted to slam the breaks on parks expansion on the grounds that management of existing lands had to be strengthened and that inclusion of questionable areas were diluting the quality of the parks system. But Congress authorized fourteen additional parks during Walker's two year tenure, and in 1976 passed the General Authorities

Act directing the Secretary of Interior to prepare annual reports identifying potential park additions. By the time President James Carter replaced Everhardt with William J. Whalen, Congress passed the Omnibus National Parks and Recreation Act (1978), which authorized fifteen additions to the park system. Among these additions, and against the opposition of the National Parks Advisory Board, was Santa Monica Mountains National Recreation Area. Shortly thereafter came the big Alaska parkland additions.[12]

The Alaska addition had its roots with director Hartzog, who succeeded in putting into the Alaska Native Claims Settlement Act of 1971, a provision requiring the Secretary of Interior to select eighty million acres he deemed "suitable for addition to or creation as units of the National Park, Forest, Wildlife Refuge, and Wild and Scenic Rivers Systems." Secretary Rogers Morton transmitted 32.3 million acres worth of recommendations for parks at a time when the existing park system comprised only thirty-one million acres. The recommendations were controversial, especially in Alaska where there was great opposition from commercial and sporting interests who feared their activities would be prohibited on much of the recommended land. It took seven years and a lot of wrangling and compromise to win over the opposition, but in 1980 Congress passed the Alaska National Interest Lands Conservation Act (ANILCA), giving the parks system more than forty-seven million acres. This exceeded the 32.3 million acres proposed by Secretary Morton seven years earlier by nearly 50 percent. The act created eight national parks, two national monuments, ten national preserves, two national historic parks, a wild river, and curtailed further expansion of the park system in Alaska by presidential proclamation.[13]

The NPS did go through a period of retrenchment during the 1980s under Whalen's successor Russell Dickenson and President Reagan's Secretary of the Interior James Watt, both of whom believed the NPS should improve its stewardship of existing parks before adding more. In 1983, the Congress supported Dickenson's Park Restoration and Improvement Program, which devoted more than a billion dollars over five years to stabilize and upgrade existing park resources and facilities. Dickenson's successor, William Penn Mott, Jr., who took over in 1989, returned the NPS to a more expansionist mode, only to have that turned around by Mott's successor, James M. Ridenour. In 1990, Ridenour, along with Secretary of Interior Manuel Lujan, Jr., did, however, preside over the largest single park donation ever: $10.5 million from the Richard King Mellon Foundation for needed lands at Antietam, Gettysburg, Fredericksburg, and Petersburg battlefields, Pecos National Historical Park, and Shenandoah National Park. Ridenour departed when Bill Clinton was elected president in 1992. His immediate successor, Roger G. Kennedy, emphasized a greater educational role for the NPS through the Internet. Kennedy was succeeded by Robert G. Stanton, who helped Secretary of Interior Bruce Babbitt kill the National Park System Reform Act proposed in Congress, which would have reassessed the need for some existing parklands and raised the standards for admitting more areas into the system in the future. Despite defeat of the bill, a consensus developed on Capitol Hill that criteria for admission into the parks system needed to be strengthened. Accordingly, Congress in 1998 amended Section 8 of the General Authorities Act to direct the Secretary of Interior to adhere to established standards in recommending sites for inclusion in the system. The amendment also prohibited the conduct of a new area study without specific authorization from the Congress.[14]

This period of the NPS also included the arrival of modern, sophisticated parks management. By the 1970s, Hartzog's rigid categorization of each park system unit as natural, historical, or recreational was causing problems. This practice did not adequately address the diversity of park areas, almost all of which contained historic or cultural resources of at least local significance. Recreational area classification implied that natural preservation would be secondary to development for heavy public use—development and use that might be ecologically harmful. Environmentalists were especially disturbed about the recreational classification of such outstanding areas as Cape Cod National Seashore and Pictured Rocks National Lakeshore. The NPS responded in 1975 by replacing its separate natural, historical, and recreational area policy manuals with a single management policy compilation addressing the range of characteristics each park possessed. A mostly natural area, for example, might also have important cultural features and portions suitable for recreational development. It would be zoned accordingly in its general management plan, and the various zones would be managed under policies tailored to each. In 1977, the old categories of park classification were abolished.[15]

In the early 2000s, NPS Director Fran Mainella presides over a park system of 388 parks in nearly every state and U.S. possession. The NPS manages these parks and supports the preservation of natural and historic places and promotes outdoor recreation outside the system though a range of grant and technical assistance programs. Major emphasis is placed on cooperation and partnerships with other governmental bodies, foundations, corporations, and other private parties to protect the parks and other significant properties and advance NPS programs.[16]

Parks on Fire Island

The first park acquired by the State of New York was the one thousand acre holding at the west end of the beach, known as Fire Island State Park—renamed the Robert Moses State Park. This occurred as with other historical events, not by direct intention to establish a park but as the by-product of a different objective.[17] The threat of a cholera epidemic introduced by the passengers aboard the *S.S. Moravia* in 1892 created the need for a safe quarantine area. The Great South Beach was selected by virtue of its isolation. With the crisis past, the state then cast about for a feasible use of the land. Thus, was created the Long Island State Park Commission. The County of Suffolk established the Smith Point Park of 345 acres to enable the mainland residents an opportunity to enjoy the recreational benefits that the area affords. Since the last remaining vacant land on the oceanfront in Suffolk was on the Great South Beach, it became obvious that additional takings needed to be made if future generations were to be provided for.

Photo 3.1. Robert Moses State Park (Aerial photo Courtesy of Long Island State Park Commission)

Photo 3.2. Robert Moses State Park (Aerial photo Courtesy of Long Island State Park Commission)

In response to the pressures for home protection, additional recreation facilities and greater conservation, various programs have been put forth. The Suffolk County Planning Commission recommended in 1960 that two miles of land at Long Cove—immediately to the east of Davis Park—be acquired.[18] Their report indicated that in addition to the desire to satisfy a recreation need, the proposal would tend to curtail further residential construction, since the park would lie in the path of eastward construction. Action was not taken on this matter. At approximately the same time the Regional Plan Association of New York recommended that all the remaining ocean frontage, including the Hamptons be acquired for public purposes. Neither of these proposals evoked much attention.[19]

In addition to the New York State and Suffolk County park holding on Fire Island, there are also seven public town beaches and the National Seashore Park with five major facilities—the Fire Island Lighthouse, Sailors Haven (home to the ancient and exotic Sunken Forest), the federal wilderness area at Watch Hill, Smith Point West, and the William Floyd Estate. This plan, in effect, was an extension of how Moses developed the Jones Beach State Park. The Fire Island Lighthouse, built in 1824 on federal land, was the first publicly owned facility on Fire Island. The second public facility was the Surf Hotel, purchased by New York State in 1892 to screen European immigrants for cholera. In 1908, Governor Charles Evans Hughes signed legislation authorizing the use of the Fire Island reservation as a state park; he appointed a Fire Island State Park Commission to manage the hotel, which it proceeded to auction off. Thus, the fear of the spread of a plague lead to the establishment of the state's first Long Island Park and the only ocean front park, until the creation of Jones Beach State Park some twenty years later? The Commission, originally made up of Edward C. Blum, Samuel L. Parish, Henry W. Sackett, John C. Robbins, and John H. Vail was underfunded and largely ignored by government until the state appointed Robert Moses to head the Long Island State Park Commission in the spring of 1924. Governor Alfred E. Smith and the state legislature created the Commission as part of a comprehensive statewide park and parkway program. The Commission, which also consisted of Townsend Scudder and Clifford L. Jackson, absorbed the meager, storm-tattered facilities of the Fire Island State Park Commission, as well as its authority over the Fire Island State Park. At that juncture, Moses began his long quest to provide public access to Fire Island, mainly through the building of connecting bridges and a connecting highway down the middle of the barrier beach. Moses's plan of 1927 for the Ocean Parkway down the length of Jones Beach included a bridge over Fire Island Inlet and an extension of the parkway continuing down the length of Fire Island. In 1938, Moses proposed to build a no-access highway on hydraulic fill twelve feet high, with pedestrian underpasses and, four drawbridges (at Fire Island Inlet, Smith Point, Westhampton, and Quogue). The proposal included the digging of a boat channel the length of Fire Island, use of the sand dredged up for highway foundation, and the establishment of three new parks.[20]

Most of these efforts went nowhere due to the nation's preoccupation with World War II and its aftermath. There was no organized public opposition to Moses's designs on the island until after the war. Before then, residents of Fire Island generally favored Moses's stewardship. In 1926, Moses established numerous public facilities, including a bathhouse that could accommodate 350 people. At the same time additional shelters were built and sanitation and water supply improvements were made. In 1928, Moses welcomed to the beach Camp Cheerful, a camp for crippled boys sponsored by the New York City Rotary Club. The camp, which provided nine cabins, an administration building, storehouse, infirmary, mess hall, helpers quarters, water supply and sanitary facilities, and equipment washed away in the 1938 hurricane, along with almost everything else on the developed part of Fire Island

Photo 3.3. Robert Moses Causeway, March 31, 1966 (Courtesy of Long Island State Park Commission)

Photo 3.4. Water Tower at Robert Moses State Park (Courtesy of Long Island State Park Commission)

just east of the lighthouse. After the storm, which was witnessed only by the state park superintendent Henry Herrnkind and his wife, plans were made to reconstruct park facilities. Because most of the protective sand dunes at the site of the old state park were washed away, a new and much more desirable site was selected about two-and-a-half miles west on higher ground. In 1940, at the new site, a modern bathhouse with refreshment and food bar, shelters, play apparatus, superintendent's and State Police headquarters, equipment shop, concrete walks, water supply and sanitary facilities, and a sheltered harbor for boats and ferries on the inlet side were open to the public. In the early summer of 1941, a stone jetty 4,800 feet long was completed at Democrat Point, at the extreme westerly end of the park. The jetty was constructed by the federal government, with cooperation from the Long Island State Parks Commission and the County of Suffolk, to arrest the westward accretion of Fire Island and to protect the new park facilities located to the east.[21]

Moses used the storm-related erosion problem to advance his grand concept of extending the public recreation system throughout the barrier beaches. Ostensibly his proposal reiterated for more than a half century that the placement of a highway would serve to anchor the dune and therefore help to forestall the ravages due to erosion.

The Moses Road Proposal

Private homeowners and commercial property owners on Fire Island reached a different and diametrically opposite view. They saw the proposed road as a larger threat to their existance on Fire Island than erosion or even storm induced demolition. If their homes collapsed due to hurricane winds, or washed out to sea, they still had the option (and the commitment) to rebuild. Conversely, if the Moses road was built many homes in the right-of-way would be irrevocably lost. They saw the issue as a battle for their very survival as property owners on the Island. Figure 3.1 depicts the Moses Plan. It clearly shows that the dune and beach to the ocean would be absent of any homes.

During the intervening years between 1938 and 1962 little public or governmental attention, other than from the USCE and homeowners, was paid to the problems of erosion on the barrier beaches. Moses, however, was diligent in his pursuit of his concept for a Fire Island ocean highway. He resurrected the issue following damage from a hurricane in 1944, and once again in 1962, shortly after a destructive northeaster storm in March of that year. This time the problems confronting the users of the Island had the interest and attention of an expanded constituency in numbers and political influence. Opposition to Moses was mounted by the property

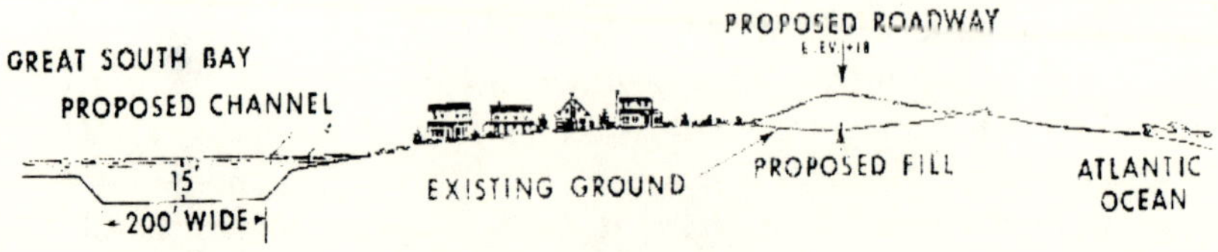

Source: Long Island Regional Planning Board, Hauppauge, NY

Figure 3.1. Moses Road Proposal

owners and renters on Fire Island, and Murray Barbash, a builder of houses on the Island at Dunewood.

Fire Island property owners consistently favored erosion control programs, even prior to the USCE study. More than 250 Fire Islanders held a meeting on August 22, 1957, attended by Congressman Stuyvesant Wainwright and representatives from the Corps, held at the Ocean Beach community. The meeting was sponsored by the Fire Island Erosion Control Committee. The impetus for the meeting was the release of a study by the USCE which they began in 1955; it called for the raising of the dunes to an elevation of twenty feet above mean high sea level, including a raised beach that would be stabilized by the construction of groins.[22]

Congressman Wainwright proposed the first effort to establish a National Seashore on Fire Island. He suggested that an eighteen mile stretch of Fire Island be designated. His recommendation for federal government involvement was in part a response to the need for erosion control as a result of the damage from hurricanes Edna and Carole in 1955. Although there was strong support from Fire Islanders for erosion control, there was an adverse reaction to the creation of a National Seashore. No action for implementation was taken on behalf of the congressman's recommendation, or the erosion control program.

The following year a committee was formed at the state level for the purpose of studying the entire issue of the barrier beach.[23] Its final report proposed that a continuous dune be constructed at an elevation of twenty feet the length of the island between the state park on the west and the county park at Smith Point.[24] The dune would then be stabilized by the construction of a "scenic" highway on the crest. The four miles of land to the east of Smith Point Park and the county lands at the Moriches inlet would be acquired through state funds at the same time, and turned over to Suffolk County for conservation purposes. At this point, the status of Fire Island became of prime concern to many persons and organizations, on and off the beach. Although the report was issued under the aegis of the then Speaker of the Assembly, Joseph F. Carlino of the city of Long Beach, it was clearly the work of Robert Moses who stridently campaigned for its implementation. He argued that the road would protect the beach from storm damage erosion and breakthrough.

Homeowners on the Island immediately recognized the threat posed by the road. Namely, the dune construction and roadway would eliminate the majority of the houses, if not all of them. In short, the entire Island from the state park at the western end to the Moriches Inlet county property would become public parkland, similar to Moses' creation of the Jones Beach State Park.

A suggestion aimed at thwarting the "Moses Plan" was offered by the opponents to the effect that the NPS be asked to create a federal seashore on the island. Hearings were held,[25] and the issue was taken up by a congressional group.[26] However, the size and scope of the park, its design and ultimate use was not resolved. There were strong doubts expressed as to the eventual implementation of this attempt.

The Move for a National Seashore

Opposition to the creation of a National Seashore was put forth by Suffolk County Executive H. Lee Dennison and the Director of the Suffolk County Planning Commission Lee E. Koppelman—albeit for different reasons.

Dennison was a professional engineer who started his career with the Suffolk County Department of Highways in 1927, the same year Moses started the development of the Jones Beach State Park. Both men became friends and both were project-oriented. Dennison viewed the Moses proposal as a logical extension of the Jones Beach concept which was recognized as one of the outstanding public recreation parks in the world. He was an ardent fan of Moses and backed most proposals Moses put forth with great enthusiasm.

Conversely, the planning director, Lee E. Koppelman, was not in favor of the road proposal. But neither was he in favor of the seashore concept as proposed. His objections arose from two concerns. The first was a conviction that barrier beaches were in a flood plain area subject to potential devastation from hurricanes and northeast storm events. Thus, houses on Fire Island would always be in danger, and Koppelman worked toward eliminating all of them. The proper environmental posture, he felt, should be to develop policy based on needs arising from the natural environment, not the needs of residents and homeowners. He also objected to the National Seashore concept because he felt it was being advanced by interests narrowly focused on defeating Moses and perpetuating the continued presence of private development.

If there were to be a National Seashore, Koppelman felt it should be of a magnitude fitting for a federal park entity. That meant that the entire thirty mile length of Fire Island should be the designated boundary in order to provide a continuous natural, public-limited, recreation area. As proposed the Seashore was more limited in size. Dennison agreed that the proposed boundaries for the Seashore were too limited, which convinced him that the thirty mile stretch of barrier island, excepting five mile long Robert Moses State Park, should be planned for and conserved by Suffolk County.

With each passing month additional construction and sale of land took place. The Suffolk County Planning Commission recommended that immediate acquisitions be made by the County, expressing their concern that by the time the federal government did act there would not be much undeveloped land to act upon. The major recommendations called for the acquisition of the four mile stretch between Smith Point County Park and Moriches Inlet County Park.[27] In addition, the two mile stretch at Long Cove was also recommended for acquisition. In the event a federal seashore was established, these lands could then be transferred on a reimbursable basis. It was also recommended that the Town of Brookhaven investigate the feasibility of initiating flood plain zoning on the remaining vacant lands between Smith Point and Long Cove. This was intended as a temporary measure to forestall development. In the event that the federal seashore would not be established, it was suggested that the land held in flood plain control be acquired by the County for conservation purposes.[28]

Powerful forces of opposition were unleashed among the residents of Fire Island and leading environmentalists. Conservationists led by Richard Pough and Curator Emeritus of Ornithology of the American Museum of Natural History Robert Cushman Murphy, successfully raised the money to purchase and protect the Sunken Forest near Point O'Woods. Fire Islanders had organized the Fire Island Erosion Control Committee, which later evolved into the Fire Island Voters Association. This group eventually put enough pressure on Governor Rockefeller, United States Secretary of the Interior Stuart Udall, and other political leaders to seriously reduce Moses' influence on the New York State Council of Parks in 1962. Moses would shortly thereafter step down from the four other state positions that he held, including the Long Island State Park Commission. Without its great champion, the idea to build a road through Fire Island languished.[29]

Other factors also intervened to spare Fire Island the construction of a highway. The experience with Ocean Parkway in Jones Beach indicated that a road would not act as an "anchor" to preserve Fire Island, as Moses had argued. The State Park Commission spent $4 million in five years to save Ocean Parkway from the ocean; the Parkway didn't save the beach.

Between 1958 and 1963, seven different bills were introduced by eight separate members of Congress, all generally supporting the concept of the establishment of a National Seashore on the south shore of Long Island at Fire Island. The one introduced on June 18, 1963 by Suffolk Congressman Otis Pike was the most specific, outlining the rough boundaries of the park that were ultimately implemented in the establishment of a Fire Island National Seashore. In September 1964, President Lyndon Baines Johnson signed the bill creating the Fire Island National Seashore extending from Smith Point County Park in the east to the Fire Island State Park in the west. Along with the mandate to protect the island from development and to preserve its unique natural features, the Seashore also took ownership and improved access to the Sunken Forest and created an eight mile wilderness area stretching over fourteen hundred acres from Smith Point west to Watch Hill, which eventually became the only federally designated wilderness area in New York State.[30]

Today, the Seashore extends from the easterly boundary of the Robert Moses State Park eastward to Moriches Inlet and includes Fire Island proper and the surrounding islands and marshlands in the Great South Bay, Bellport Bay, and Moriches Bay adjacent to Fire Island; Sexton Island, West Fire and East Fire Islands, Hollins Island, Ridge Island, Pelican Island, Pattersquash Island, and Reeves Island. In addition, the waters surrounding the Seashore includes up to distances of one thousand feet in the Atlantic Ocean and up to four thousand feet in Great South Bay and Moriches Bay.

Photo 3.5. Robert Moses State Park (Courtesy of Long Island State Park Commission)

The first mention of creating a National Seashore on Fire Island was in a NPS report released in 1955.[31] During the 87th Congress (1963), representatives John Lindsay and William Ryan of Manhattan introduced bills to carry out the recommendations of the report. Active congressional consideration of a proposed seashore was intensified that year.

Secretary Udall was supportive of the 1955 NPS recommendation for the Seashore, but expressed his opinion in June 1962 that the boundary had to extend to the westerly boundary of the Village of Southampton—a total distance of fifty two miles containing eight thousand acres—instead of terminating at Moriches Inlet. Representative John Lindsay of Manhattan introduced H.R. 3693 that generally conformed to Mr. Udall's position.[32] The issue is that a National Seashore is to serve the American public and therefore should be of sufficient size to accommodate more than local users. His bill however, stipulated that the park not exceed 7,500 acres. It further provided that as soon as the first five hundred acres were acquired, the Secretary of the Interior shall declare the establishment of the National Seashore.

Reactions to the bill were immediate and reflected strong positive and negative responses. Fire Island residents and active nonresidential environmentalists welcomed the opportunity represented by the Seashore, albeit not for the same reasons. Fire Islanders saw the measure as the death knell of the Moses road and the increased potential for governmental support for erosion control. After all, if a national park were established, it followed that government would prevent its destruction from the forces of erosion. Environmentalists saw the bill as the vehicle to stop development and lead to enhanced conservation. The Regional Plan Association of New York interpreted the bill as an important step in the implementation of their recommendation to preserve the entire undeveloped south shore ocean frontage from Jamaica Bay to Montauk.[33]

Opposition to the bill arose from two quarters. The residents on Dune Road east of the Moriches Inlet and the Board members of the Town of Southampton were not opposed to a National Seashore on Fire Island, but were vehemently against Udall's extension of the park to include their community. The other opponents were Representative Otis Pike and County Executive Dennison. Both Pike and Dennison opposed a National Seashore on any of Suffolk's barrier beaches. Dennison expressed his preference that Suffolk County government buy the undeveloped land. This also reflected the position of the Suffolk County Planning Commission.[34] In Dennison's weekly radio talk of Tuesday, April 1, 1963, he said: "If we are going to talk about preserving Fire Island . . . for natural-state wildlife and splendid isolation, then let us talk about it as such, rather than from the efforts and viewpoints of a handful of people who started the whole thing mainly to prevent the construction of an ocean boulevard along the barrier beach. If we are going to conserve Fire Island, then let us here at home conserve it."[35]

Representative Otis Pike joined the debate, claiming that a discussion between himself and the Secretary in 1962 led him to conclude that Udall felt the seashore proposal was dead. This was immediately countered by the Secretary, who disputed Pike's claim, stating that his comments were only an expression of his concern that Congress at that time was not showing great interest.[36]

Senators Kenneth B. Keating and Jacob Javits were attuned to the ferment over the future of Fire Island and were thinking of introducing a companion bill. Senator Javits was aware of Dennison and Pike's opposition to the Seashore from a memo he received from Richard Aurelio of his staff. Javits wrote to Governor Nelson Rockefeller to solicit the Governor's opinion of the Lindsay bill. The response on behalf of the Governor came from Dr. William Ronan, a close advisor and confidant of Rockefeller, who advised the Senator not to get involved. Lindsay's support of Udall's desire for an expanded park included too much acreage and the opposition from Southampton and Suffolk officials was too controversial, Ronan felt.[37]

Representative Ryan of Manhattan then introduced his bill in an effort to ameliorate the conflicts generated by Lindsay's bill. He proposed that the boundary be limited to a distance of two miles east of the Moriches Inlet. Since this property was already Suffolk County parkland, no homeowners on Dune Road would be affected.[38]

His bill was then followed by Senators Keating and Javits who jointly introduced their bill which mirrored the House bill of John Lindsay.[39]

Both Senators received a flurry of letters, mainly from New York City residents who also had property on Fire Island and/or Westhampton Beach. Those who had homes on Fire Island, such as the noted radio anchor Charles Collingwood and Gardner D. Stout, were strongly in favor of the Senate bill. Those whose property was east of the Moriches Inlet, including the Dune Road Property Owners of Westhampton Beach, wrote in strong opposition to the Udall proposal to include their community.[40]

One result of the attention paid to Fire Island by the media was a shift in position by the Suffolk County Board of Supervisors which initially was opposed to supporting erosion control expenditures on Fire Island. They now took a favorable position on the creation of a National Seashore.[41] Several factors account for their change of heart. First, they recognized a strong vocal citizen rally behind the concept. Second, while they still refused to appropriate the County's share of $3 million for the USCE erosion plan, they voted unanimously to support the park with the explicitly stated expectation that the federal government would now pay the entire cost of erosion control. Third, Pike and

Dennison were Democrats, while the Board was predominantly Republican. However, their stance was not totally that of the "loyal opposition." Dennison and Pike wanted the County to buy all of the privately held undeveloped land on Fire Island. The Board preferred to transfer such costs to the feds.[42]

During the following months, nine more bills were introduced in the House of Representatives. They generally fell on the side of supporting Udall's call for a fifty-two mile Seashore, or on the side of a truncated boundary limit to either the west side of Moriches Inlet, or not more than $1^1/_2$ miles east of the inlet.[43]

One week after the Board of Supervisors came out in favor of the Seashore, Representative Pike changed his stance and came out in favor of the park, but with limits on the Udall plan.[44] On June 18, 1963, he introduced his bill, which became the generally accepted approach in the House.[45] It supported the limitation of the eastern boundary to the Moriches Inlet, thus removing the opposition of the Westhampton Beach opponents, and Speaker of the New York State Assembly Perry B. Duryea, and then President of the Long Island State Park Commission.

The Pike bill excluded the existing communities on Fire Island from the general threat of condemnation, with the exception of an eight mile reach from Ocean Ridge to the western boundary of the Smith Point County Park, in which the NPS would have the right to condemn any private property in accord with three alternative provisions:

1. Property to be condemned must be a fair market purchase taking; or
2. Owner can retain life tenancy. The final price to be reduced from full market value by the value of the lifetime use; or
3. Owner has a twenty-five year ownership.

The powers held by the NPS provide that no zoning actions by the towns and villages on Fire Island will be approved if they are inconsistent with the Act. Furthermore, if any owner violates the provisions, the Secretary can exercise condemnation. Within the eight-mile wilderness of the Seashore, from Smith's Point County Park to Watch Hill designated in 1980, access can only be by ferries and footpaths.[46] The only road usage is for park maintenance. Pike's bill also provided that user fees may be collected, and the Secretary may undertake erosion control on Fire Island in so far as such measures are consistent with the purposes of the Act. This last proviso, which satisfied the Fire Island property owners at the time, contained disparities in what erosion techniques could be used, depending on whether the goal is to minimize perturbations of the natural setting, or maximize property stabilization.

The bill attempted to provide citizen input by creating a Fire Island Advisory Commission composed of fifteen members who would serve unsalaried two year terms. These members would be appointed by the Secretary of the Interior. Each town board in Suffolk County would recommend a member. Brookhaven and Islip would each receive one additional member in recognition that the entire Seashore lies within those two towns. The Governor of New York State, County Executive of Suffolk County, and the Secretary would each recommend one member. Thus, the initial step in the Federal legislative procedure was accomplished.

This, however, never shook Robert Moses's commitment to his belief that the road proposal was the only intelligent plan for the Seashore.[47] Although there was a consensus shared by all the players in the creation of the Seashore that the road issue was dead, particularly in view of Governor Nelson Rockefeller's declaration early in 1963,[48] Moses issued a memorandum in 1963 stating a dozen questions regarding Fire Island's future. His preamble to the questions and his answer to the questions all hinged on his road proposal.

> The future of Fire Island is being debated with more heat than light. I believe I have had long experience as a public official, resident and year-round enthusiast in Atlantic shore front preservation in the State of New York. For forty years I have been engaged actively in acquiring and preserving the beaches and making them accessible in as nearly as possible their natural state. I live most of the year in Suffolk and have been a year-round sailor and boatman, swimmer, fisherman, clam digger, gatherer of wild oysters, scalloper, bird watcher, everything in sport but a hunter and gunner.[49]

His three major conclusions were that there should be a National Seashore on Fire Island, a concession to his opponents. But he suggested that there should be a two-lane roadway to protect the barrier beach, stating, "I believe no other (plan) is feasible."[50] In his strongly stated opposition to Udall's suggestion for a low-rise, inexpensive perched beach sand fill to cope with erosion, Moses stressed that only the continuous high hydraulic sand fill dune structure on which the road would sit was sufficient.

One decade after the establishment of the Seashore, Moses once again tried to gain approval for his road proposal. In correspondence with the Seashore Superintendent James Godbolt, Moses requested a meeting and asked what policy, if any, the Seashore had in regard to the extension of the ocean highway.[51] Godbolt responded that he would welcome the opportunity to meet with Moses, but he was opposed to allowing a road through the entire Seashore as it was contrary to the purposes of the park.[52] Moses's response to Godbolt was not designed to win any friends. He stated: "Your logic entirely escapes me and bears no logical relation to our long experience here. You are going to come a hell of a cropper if you attempt to use such gobbledegook with the endorsement of the Department of the Interior."[53]

The Legislative Process

The introduction of bills in the House or Senate is only the first step in the process of adoption. Both Houses of Congress follow a structured committee system. Prior to any legislation reaching the floor for consideration, it must first be referred to the prime and related committees that exercise control over matters within their substantial jurisdiction. The prime committee over proposed national parks in both Houses is the Committee on Interior and Insular Affairs. Within the overall committee there may exist subcommittees. The applicable subcommittee in this instance in the House, for example, is the Subcommittee on National Parks. Each of these entities may schedule public hearings on any proposed legislation submitted to them. If the pending matter has any controversial issues (and most bills do), the hearing process may be a lengthy one. Even when the general consensus throughout the Congress is positive, the timing from introduction to approval can take more than one year due to the heavy agenda of the Congress. Further delays occur when amendments are proposed in the subcommittees, the full committee, or the entire House and/or Senate. Since the House and Senate versions of the proposed legislation may differ, there must be conference discussions to arrive at a unified final bill. Assuming the bill is not killed at any step in this iterative process, the adopted legislation must then be submitted to the president for signing.

Thus, it is imperative for any congressman who introduces a proposed bill to request expeditious handling of the hearing process. For example, Senators Keating and Javits wrote to Chairman of the Interior and Insular Affairs Committee, Senator Henry "Scoop" Jackson, mentioning their introduction of S.1365, and requesting that hearings be expeditiously scheduled. This was sent at the end of April 1963. Senator Jackson sent his reply to Senator Javits on December 6, 1963, to inform him that the hearing on his bill would be convened on December 11, 1963.[54] Throughout all the bill introductions and debates over the conflicting issues of proposed boundaries, Secretary Udall held firm to his view that the Seashore should extend to the Village of Southampton's western boundary. This conflict between Udall and the Town of Southampton and the residents of Westhampton Beach could easily have been a "deal killer." At the last moment Under Secretary of the Interior James K. Carr testified at the hearing that they submitted a substitute bill. Although it was still their professional view that their preferred choice was for a larger Seashore—from the west end of Fire Island all the way to the western boundary of the Village of Southampton—they could live with a truncated version that would terminate 1½ miles east of the Moriches Inlet. This capitulation reduced Interior's June 10, 1963 proposal for an 8,000 acre park to one approximately 5,700 acres, of which 3,500 acres were to remain unimproved with no road. Leonard Baker of *Newsday* noted that Interior's position on roads presented Moses with a dubious accolade as "the real father of the proposal for a Fire Island National Seashore" due to his road proposal.[55]

Subsequent to the establishment of the Seashore in September 1964, another seemingly unrelated event culminated in the addition of the William Floyd Estate to the holdings of the Seashore.

General William Floyd

The adoption by the Suffolk County Planning Commission of Suffolk County's first open space plan of 1960 included a recommendation for the acquisition of the Weld Estate in the Town of Smithtown.[56] This was one element in the proposal to preserve the four major river watersheds in the County. The Weld property contained the major headwaters of the Nissequogue River. The County's Comprehensive Open Space Plan released in May 1964 repeated the recommendation for acquisition. Robert Moses responded to the plan by informing Suffolk Planning Director Lee Koppelman that he would provide $2.5 million towards the acquisition, and with that commitment, Koppelman met with Weld and they worked out a deal for County acquisition. Weld was a dedicated environmentalist and a publicly involved citizen.[57]

Although Weld was a principal in the Wall Street firm of White Weld and Co., he did not possess enough wealth to allow for a total donation. Weld informed Koppelman that in addition to the Smithtown property, his wife Molly (nee Floyd) was the executrix of the Floyd Estate and that he could afford to sell one of the properties, either the Smithtown one or the Floyd Estate, at a reduced market value price and also donate the other property.[58] He asked the planning director for his choice. The planner observed that the property in Smithtown was part of a plan to acquire interests in the entire river valley and a money commitment from the State was available. Conversely, the Floyd Estate was of national significance, since Floyd was the only Suffolk County signer of the Declaration of Independence. In addition, the Floyd Estate was contiguous to the existing United States Fish and Wildlife Sanctuary—the Wertheim National Wildlife Refuge. In view of the creation of the nearby Fire Island National Seashore, Koppelman suggested that it would be more appropriate for the Floyd Estate to become part of the Seashore. Weld concurred and the County proceeded to acquire the property in Smithtown at a reduced market rate. Congressman Pike was informed of the offer of donation by Mrs. Molly Weld and he submitted the necessary legislation.

Pike introduced his bill on May 10, 1965, calling for the acceptance of the donation by the Department of the Interior.[59] Weld also wrote to New York State Assembly Speaker Perry Duryea, Jr., asking for his help to get a companion bill introduced in the Senate. On June 18, 1965, Duryea wrote to Javits asking for the Senator to co-sponsor the bill with Senator Robert Kennedy. On June 23, 1965, Patricia Connell, the legislative counsel to Javits responded that the Senator already had his colleague Kennedy as a co-sponsor. The Senate bill authorized the Secretary to accept the donation of 611 acres. On August 16, 1965, Assistant Director for Legislative Reference from the President's Bureau of the Budget, Phillip S. Hughes, wrote to Senator Henry M. Jackson, chairman of the committee on Interior and Insular Affairs, indicating the president's support for the acceptance of the donation.[60]

The House approved the Pike bill on October 9, 1965, followed by Senate action supporting H.R. 8035 with no amendments. Thus, with relative speed this significant property of important environmental value was added to the scope of the Seashore.[61]

Chapter 4

Historic and Cultural Resources

An historic resources study of Fire Island completed in 1979 identified additional cultural resources. In all, the study identified fifteen zones in the Seashore with historically significant resources, including lifesaving stations, the fish factory, Camp Cheerful, historic homes, and more. An archeological reconnaissance in 1974 found no premodern remains on Fire Island, but there is continuing interest in underwater archeological surveys that many feel would reveal important maritime artifacts.[1]

The Seashore manages two major cultural resources—The William Floyd Estate and the Fire Island Lighthouse.

Photo 4.1. Revolutionary War General William Floyd (Courtesy of National Park Service)

Photo 4.2. William Floyd Estate (Courtesy of National Park Service)

William Floyd Estate. The William Floyd Estate is an historic mainland property north of the Moriches Bay near the eastern end of Fire Island in Mastic Beach. The rambling former residence of General William Floyd, one of the four signers of the Declaration of Independence from New York, contains a fully furnished twenty-five room house, originally built in the 1720s (but with nineteenth- and twentieth-century additions), ten farm outbuildings, a family and a slave cemetery, and roads, trails, gardens, fields, and other sites of minor historical significance. The estate dates from the original patent to Colonel William Smith by the Province of New York in 1693 and had been in the ownership of the Floyd family since 1718. At the time of the first General Management Plan in 1976, public use of the estate was negligible, and maintenance of the cemetery and occasional patrols were the only federal management efforts. The residence building was occupied by Mrs. John T. Nichols—a direct descendant of General Floyd—under a twenty-five year lease agreement with the NPS. In October 1976, the Floyd descendants relinquished their use and occupancy rights to 33.8 acres of the estate, including the major residence and outbuildings. The remainder of the estate was used by the family as a private hunting preserve until November 1991, when the twenty-five year lease expired. The estate was entered in the National Register of Historic Places on October 15, 1980. In 1980, the Seashore made the William Floyd Estate a separate unit of the Seashore, with a unit manager reporting directly to the Superintendent. Before that, the park historian Steve Kesselman, had reported to Neil Bullington, the Chief of Interpretation (see organization charts in organization chapter).[2]

Kesselman, the Floyd unit manager from 1976 to 1984, has called the estate the "perfect historic site," based on the Seashore's decision to interpret the Floyd estate on an "historical continuum" and not as an archeological object dating from a certain point in history. "The idea was to discuss historical aspects of the estate in terms of how they evolved over time, as they were acquired," Kesselman explains. This decision caused some trepidation among some Floyd family members, who feared the Seashore's interpretation might impinge on the family's privacy. But after assurances, the estate opened in 1981 to the public with praise from the Floyd family's primary donor, Ms. Mary(Molly) Weld. "Every room I go into," Kesselman recalls Mrs. Weld telling him, "I think I'm going to see my mother [Cornelia Floyd Nichols]."[3]

The first summer only the first floor of the main house was opened to the public and received about nine thousand visitors. By the second summer the second floor had opened to the public as well. Construction was completed on a parking lot and associated road system in 1987, which opened up the estate to higher levels of visitor use. That same year, a pump house to pump potable water to the manor house and new parking area was also built at the Floyd Estate and 2.5 miles of previously unusable trails at the William Floyd Estate were opened with use of recently purchased bomford flail mower. In 1989 a "fire break"—basically, a clearing of flammable brush—was constructed around part of the estate to address a high incidence of arson. Also in 1989 fifteen of eighteen historic fields at the William Floyd Estate were reclaimed, formal gardens within the Estate's historic zone were upgraded, and older, less productive perennial plantings were replaced. Visitors were enabled to enjoy the visual experience of open areas and maintained landscapes similar to that of historic times.

A lack of environmental controls (i.e., temperature, humidity, light ultraviolet radiation) has resulted in damages to structural elements, museum objects, and the historic fabric in the estate main house. For instance, an infestation of the Powderpost Beetle, *Annobiid sp.*, has been identified as present in the house and furnishings since the Floyd family occupied the site. A monitoring program, developed in 1984, continues with the intent of capturing, identifying, and documenting the insects when they emerge. For several years, objects suspected of having active beetle infestation were fumigated with CO_2 by the NPS North Atlantic Regional Office conservation staff.

The estate staff have also managed an ongoing herbarium collection with several hundred species, including a large box turtle inventory. Recaptures of box turtles going back to the 1930s have been documented still roaming around the estate grounds. In 1987, the Museum of Natural History duplicated the original J. T. Nichols journals at no cost to the NPS, so that researchers will no longer have to use, and damage, the original journals. In 1988 over 190 documents dating back to the 1680s were donated to the estate by a distant relative of the Nicholls family. In 1993, environmental protection of the estate was upgraded by installation of new oil and gas furnaces in the main house. The year 1995 was an important one for the William Floyd Estate. In that year a Memorandum of Agreement was signed with Friends of the William Floyd Estate for that group to raise funds to improve visitor facilities at the estate. The amount of money ultimately raised for the Estate by the Friends in subsequent years was negligible and the relationship has since lapsed. In 1995, as well, the installation of storm windows, a physical security system, an upgraded electrical panel, and upgraded intrusion and fire systems were completed. Also in 1995, a Cultural Landscape Assessment of the William Floyd Estate was completed, as was the first phase of construction of employee housing and a curatorial storage facility. In 1996, the staff residential quarters were completed, at a cost of roughly $250,000. Today a Park Protection Ranger and a park biologist occupy these quarters. In 1998, the curatorial storage facility was completed, at a cost of roughly $250,000. This facility houses park archeological, historical, natural historical, and archival holdings. The park collections include over 27,000 historic objects, 80,000 pieces of archival material, 28,000 archeological objects and approximately 2,500 natural history specimens. Many objects of cultural interest are used in exhibits, the majority at the William Floyd Main House. The remainder are in storage in the Resources Management Offices and the William Floyd Estate.[4]

The William Floyd District is presently managed in accordance with the site's Development Concept Plan (Description of Selected Alternatives), and the Cultural Landscape Plan.[5]

Fire Island Lighthouse. One of the areas of the General Management Plan that the Seashore acted on almost immediately was the preservation of the Fire Island Lighthouse, acquiring it from the United States Coast Guard in 1978. The Fire Island Lighthouse is comprised of eighty-two acres and has been an important landmark for transatlantic ships coming into New York Harbor at the turn of the last century. For many European immigrants, the Fire Island Light was the first sight of land upon arrival in America. The first lighthouse built on Fire Island was completed in 1826. It was a seventy-four foot high, cream colored, octagonal pyramid made of Connecticut River blue split stone. The tower was built at the end of the island, adjacent to the inlet. This tower's low vertical height rendered it ineffective. It was taken down and the stone was reused to build the terrace for the present lighthouse. Today a circular ring of bricks and stone are all that remain of the original lighthouse. Due to the westward migration of sand along the beach the inlet is now approximately six miles westward of this site.[6]

Photo 4.3. Fire Island National Seashore Lighthouse (Courtesy of Fire Island National Park Service)

Photo 4.4. Fire Island National Seashore Lighthouse (Courtesy of Fire Island National Park Service)

In 1857, Congress appropriated $40,000 for the construction of a new 168 foot tower. It was lit for the first time on November 1, 1858. This tower was made of red brick and painted a creamy yellow color. The tower was changed to the present day mark of alternating black and white bands in August 1891.

The new tower was fitted with a First Order Fresnel Lens, which emitted a white flash at one minute intervals. A Funk Lamp with four concentric wicks was used for illumination. Over the years, various fuels were used for the lamps, including whale oil, lard oil, mineral oil, and kerosene. Electricity finally reached the lighthouse on September 20, 1938. However, on September 21, 1938 a hurricane struck the island, effectively severing all electric power to the island and causing a delay in the electrification of the Fire Island Light Station.

The United States Coast Guard has been present on Fire Island since its inception in 1915. A Coast Guard Station was established on the Lighthouse tract. Eventually the United States Lighthouse Service was dissolved. The administration of lighthouses was placed under the jurisdiction of the U.S. Coast Guard in 1939 "in the interest of economy and efficiency" (Presidential Reorganization Act).

Photo 4.5. Fire Island Lighthouse (Courtesy of Titus Kana, www.FireIslandVision.com)

For the first 137 years of the Fire Island Light's existence, the Lighthouse station was accessible only by boat, by foot, or ice skates if the bay was frozen. On June 13, 1964 that changed when a bridge was built over the Fire Island Inlet to provide access to newly christened Robert Moses State Park. Nearly 150 years of light keeping on Fire Island ended on December 31, 1973 when the lighthouse was replaced by a strobe light on the nearby Robert Moses State Park water tower. After the decomissioning of the Fire Island Light, the Coast Guard gave the NPS a five-year permit to use the entire Lighthouse Tract. Between 1974 and 1980, private citizens grouped together in an effort to "save the Fire Island Lighthouse." The strobe light on the Robert Moses Tower only shone seaward and was of no use to boaters on the Great South Bay. Public support for restoration of the Fire Island Lighthouse was great among the Baymen. In 1979, the tract was declared by law to be within the boundaries of the Fire Island National Seashore. With limited funds, the major function of the NPS during its early administration of the Lighthouse tract was to prevent further deterioration of the buildings from neglect and vandalism.

In 1982, the Fire Island Lighthouse Preservation Society was formed (FILPS). This group successfully raised over $1.3 million for the restoration and preservation of the Fire Island Lighthouse. Since its inception, FILPS's directive has been to work with the NPS to help preserve the nautical heritage of Fire Island and Long Island, and to ensure that the monument remains an integral part of that heritage.

In 1984, the Fire Island Lighthouse was placed on the National Register of Historic Places. The purpose of this designation was to preserve and interpret the historic resources of the lighthouse complex, as well as to manage natural resources within the lighthouse tract and the adjacent bay islands for interpretation, environmental education, research, and preservation. The 1858 Lighthouse and keeper's quarters, radio transmission station, and foundation of the 1826 Lighthouse are significant for their role in the history of coastal navigation, in the development of private maritime industry in bay and ocean, in symbolizing changes in federal maritime responsibility, and in the naval role in American wars.

The Lighthouse was restored to its 1938 condition, which is when electricity was first installed. In preparation for its relighting in 1986 the NPS constructed a universal access walk at Lighthouse Beach with donated materials and labor provided by Friends of Lighthouse Beach. A Commercial Use License was issued to Captree Boatman's Association authorizing charter ferry service to the Fire Island Lighthouse Tract, and boardwalks were completed at the Lighthouse Tract. The maintenance staff completed renovation work to the electrical system and did structural repairs of Lighthouse for relighting of Fire Island Light on May 25. On Memorial Day, May 28, 1986, the Fire Island Lighthouse was relit and reinstated as an official aid to navigation.

In 1987, an external recoating and painting was completed at the Fire Island Lighthouse, completing Phase III of a four-phase project. The FILPS raised funds for restoration work at the Lighthouse. Fire Island residents donated funds to acquire a replacement house for the Lighthouse area. An agreement was executed for emergency ambulance service to the Fire Island Lighthouse Tract by the Village of Babylon Fire Department. In 1988, $5,500 was made available from the FILPS to help keep the Keepers Quarters Museum open during the summer. An additional $4,500 was provided by the same group for increasing maintenance levels. On February 15, 1989, WLIM signed a historic lease for use of the Lighthouse tower and a $100,000 check was presented to the NPS, which constituted the first two years of lease payments. Leases for the next thirteen years were to be paid to the NPS at a rate of $36,500 annually. In 1989, a crushed concrete surface was added to the Lighthouse road. The Celebration of the Bicentennial of American Lighthouses included the installation of handicapped access

Photo 4.6. View of Western Fire Island from Fire Island Lighthouse (Wikipedia Encyclopedia)

ramps to the Lighthouse Keepers Quarters and Tower, and renovations to the tower facility to allow visitor access to the gallery room. These projects were made available due to a donation from the FILPS.

By 1993, the Seashore had complained that they had received only minor budget increases, all of which had fallen short of meeting the increased costs of operation. This resulted in the closure of the Fire Island Lighthouse, lapsing of positions, and curtailment of overtime, travel, and training. It also spurred negotiations with the FILPS to take over the operation of the Lighthouse.[7] In December 1996 the FILPS, through an agreement with the NPS, took over the maintenance and operation of the Fire Island Lighthouse and Keeper's Quarters. Today the light is lit by two 1000-watt bulbs, which rotate in a counterclockwise direction, giving the appearance of a flash every 7.5 seconds. The light is visible for approximately twenty-three miles.

The FILPS continues to work with the NPS, which owns the lighthouse and surrounding property, and the Coast Guard, which maintains the Lighthouse's optic, to allow the Fire Island Lighthouse to serve the public. Not only does the lighthouse guide mariners in the Great South Bay and Atlantic Ocean as it had for years, but it now also serves as an educational center. The keeper's quarters contain a museum exploring the history of the Lighthouse and the U.S. Life-Saving Service. The boardwalks around the Lighthouse and surrounding dunes allow visitors to enjoy the unique marine environment. Society volunteers and staff conduct tower tours that give visitors the opportunity to walk around the gallery—Fire Island is the only lighthouse on Long Island that allows this—while discussing local history, Lighthouse history, and Life-Saving Service history. The Fire Island Lighthouse is a popular destination, with about one hundred thousand visitors each year.[8]

Smith Point West. In 1965, important negotiations were held with Suffolk County Executive, H. Lee Dennison in an effort to obtain a lease of a portion of County land west of Smith Point County Park for use as a visitor center site. In July, the County Board of Supervisors approved the construction of the Smith Point Visitor Contact Station by the NPS. Actual construction of three prefabricated geodesic dome coverings began on July 22, 1965. One dome was used as an exhibit unit, one as a visitor information center, and one for picnicking. These domes have since been removed and there is today only a small checkpoint station. Smith Point is open daily, year-round and the visitor contact station serves as a checkpoint for vehicle beach access in the spring and fall. Although accessible by automobile, there is no parking at this facility. Visitors must park at Smith Point County Park and walk over to the Seashore facility. Smith Point is the entry point to the Fire Island Wilderness, which is used by sportsmen for fishing and waterfowl hunting during the New York State hunting and fishing seasons. Most of the area between Watch Hill and Smith Point is designated and managed as a cooperative hunting area. Motorized vehicles are not permitted in the designated wilderness, but a limited number of seasonal sportsmen's vehicle permits allow access to the beach in front of the wilderness area from Smith Point to Long Cove, from September 15 through December 31 only. Although this portion of the beach is used extensively by bathers in the summer months, there are no park service lifeguards, and therefore the area is not designated as a swimming beach.[9]

High Dune Wilderness Area. Section 7 of the enabling law designating the Sunken Forest Preserve and the federal lands east of Davis Park, known as the eight-mile zone, as natural areas to be preserved in as nearly their present state as possible. Access to the eight-mile zone is limited by the enabling legislation to footpaths and ferries. The eight-mile zone is of particular importance because it is predominantly undeveloped. Most of the saltmarsh occurring on Fire Island is found in this area. Also located here are the larger mammals, including but not limited to white-tailed deer, red fox, cottontail rabbit, and muskrat. The preservation of both the Eight Mile Zone and the Sunken Forest in their natural states is an explicit mandate of the park's enabling law.[10]

Within this zone the Watch Hill area was developed to accommodate ferry and boat access to this portion of the Seashore. Located at the extreme westerly end of the natural zone a short distance from the exempt community of Davis Park, the development has been criticized

Photo 4.7. Smith Point County Park (Courtesy Long Island Regional Planning Board)

by some as being too elaborate. This is one of the last remaining segments of natural barrier island along the North Atlantic coast, and the only one designated as such in New York State. Driving in the natural zone had been a continuous source of controversy in preceding years, resulting in a request by a local conservation organization that the natural zone's potential for wilderness classification be investigated during the General Management Plan process. An investigation to determine the feasibility of wilderness designation was completed in the fall of 1979, and on December 23, 1980, Public Law 96–585 was enacted by Congress designating approximately 1,363 acres of the park as the Otis Pike Fire Island High Dune Wilderness Area, making the Seashore subject to the federal Wilderness Act of 1964. The purpose of this designation was to preserve the wilderness character of this area. All land within this area, with the exception of Bellport Beach, is in Seashore ownership. Residential use of twenty-one federally owned residences under terms of reserved use and occupancy extended through 1992. On September 30, 1993 all use and occupancy rights on these properties expired, leaving only one life estate, which has also since expired. All buildings have been removed and these parcels have been allowed to return to their natural state.[11]

Old Inlet Area. There were originally two private clubs situated very close to each other in the eight-mile zone. One was the Old Inlet Club and the other was Fireplace Beach Club. Fireplace Beach Club consisted of 1.04 acres bay to ocean; Old Inlet consisted of 3.2 acres bay to ocean. Since the Seashore had no immediate plans for these areas when they were acquired in 1968, at the request of the membership of the two clubs and for a fee, the Seashore allowed the members to continue to use these clubs through the use of special use permits. These special use permits were not renewed in 1974. The facilities at Fireplace Beach Club were removed but some of the facilities at Old Inlet were maintained. Today Old Inlet contains a twenty-six-slip boat dock, seasonal public restrooms and beach access. Old Inlet is one of two visitor areas in the Seashore that does not have visitor contact facilities (see Talisman below). Both are seasonal, day-use only visitor areas, accessible by private boats only. However, Old Inlet is a two-mile hike from Smith Point. The Friends of Fire Island National Seashore, a nonprofit organization established in 2000 to advocate for more support for the park, saw one of its projects materialize in 2003 with the installation of the Old Inlet Comfort Station, an environmentally friendly, solar powered composting toilet. [12]

Watch Hill. This is the largest developed area in the Seashore and lies within the Wilderness Area. The Act that established the Seashore gave the Secretary of the Interior the right to condemn all lands within this zone which extends from Smith Point West to the eastern boundary of Davis Park. A small section of the west end of this area was developed by a builder and called "Bayberry Dunes." The federal government acquired these houses and at the time carried on negotiations with owners. Most of the owners were given the right to stay in these houses for a certain number of years which varied from five to twenty-five years or life occupancy. No one chose to stay beyond May 1978, which is when the Seashore acquired the last building.[13]

The marina was completed in July 1967. The Seashore operated the 166 boat marina at Watch Hill until 1969 when it was given to a concessioner. In 1970, a concession building was completed. On July 11, 1973 a facility with an office, exhibit area, first aid and locker room, marine storage, restrooms and showers, grocery store and marine shop, and lifeguard locker room were completed. In 1970, the State of New York passed a law requiring contained heads on all boats. A pump-out station was installed at the Watch Hill Marina in 1972 and the law was enforced at both Sailors Haven and Watch Hill Marina in the summer of that year.

In 1989, Watch Hill Visitors Center was rehabilitated through Harpers Ferry Center. In 1993, the park was funded for planning and design of a Watch Hill ferry terminal center, which was to be developed in conjunction with the Village of Patchogue River Redevelopment Plan. Plans for the visitor center also included an Education Center in cooperation with the Sea Grant Program of Stony Brook University. In 1993, a Youth Conservation Corp crew constructed a helicopter landing pad

Photo 4.8. Watch Hill Nature Walk (Courtesy of Titus Kana, www.FireIsland Vision.com)

at Watch Hill to service the Watch Hill area and neighboring community of Davis Park.[14]

Today, Watch Hill includes a 150-slip electrified marina, ferry service, campground (with a handicap accessible site), visitor center, and a lifeguarded beach. The visitors center/marina area contains a full service restaurant, pizza shop, snack bar, and general store with souvenirs. Watch Hill has approximately three miles of wooden walk, and approximately eighteen hundred linear feet (0.34 miles) of wooden walk from Watch Hill to the Davis Park boundary to the west. A dune station/visitors center contains the restrooms/ showers facility, the lifeguards' office and First Aid Station. There is a horse barn and paddock as well. Watch Hill contains a tent campground area. There are twenty-five family sites and a group campsite which will accommodate up to sixty people. Watch Hill is open from May 15 through October 15. Visitor activities include camping, beach activities, and ranger-led activities such as canoeing and marsh walks.[15]

Talisman/Barrett Beach Area. Talisman was acquired in 1965. The area's motel units, restaurant and bar units, and houses were developed originally to serve members of the United Nations and was frequented by famous celebrities such as Carol Channing, Hugh O'Brien, Hugh Hefner, Bill Cullen, and relatives of famous political figures. The motel units have since been converted to apartments for the house seasonal employees of the Seashore. Talisman is one of two visitor areas on Fire Island that do not have visitor contact facilities, the other one being Old Inlet. Both are seasonal, day-use visitor areas, accessible only by private boats. However, Talisman was once reachable by ferry to Barrett Beach during the summer season. But Barrett Beach, which was donated to the Seashore by the Town of Islip in 1998, has been without regular ferry service, lifeguards, and other amenities since the end of that season; all facilities were closed for safety reasons. The Seashore received funding in 2001 to remove the deteriorated marina, restrooms, and concession building, and to construct replacement facilities. The Talisman restroom and picnic area were improved in 2002. A new concession and maintenance building was completed in 2003. The ferry loading dock and a dock for visitors to off-load private boats is nearing completion.[16]

Sailor's Haven/Sunken Forest. Sailor's Haven was acquired on April 14, 1966. It is one of the developed areas of the Seashore and used to be a private yacht club. It consists of 36.06 acres of land and is the only place on the island where the Seashore owns the lands under water—about 1,000 feet in a northerly direction and 1,371 feet in an easterly-westerly direction. Although $558,000 was set aside for this acquisition, when the case came to court, the owners were awarded $375,000 as a fair price for this property. Since its acquisition, boardwalks throughout the Sunken Forest have been built, along with rest areas within the forest and boardwalks leading to the beach. The Sunken Forest trail has 6,500 linear feet (1.23 miles) of wooden walk and 2,900 linear feet (0.55 miles) of concrete walk. There is a staff housing unit and a horse barn. Sailor's Haven facilities include a 46-slip electrified marina and anchoring-off area for many more boats, a snack bar, a souvenir shop, showers, and a visitor center.[17] The Sunken Forest was acquired in May 1966. In 1964, a meeting was held with President of the Sunken Forest Preserve, Inc., James N. Dunlop, to discuss donation to the NPS of the seventy-five acres of the Sunken Forest property that the group held east of Point O'Woods. Under the terms of the donation the NPS was required to control access to the area by erecting a cyclone fence with gates. The Sunken Forest Preserve donated $3,500 for this purpose with the NPS paying the balance of the costs. Sailor's Haven and the Sunken Forest are managed as one unit.[18]

Visitation. In 1986, visitation to NPS administered areas was 613,824, a decrease of under 3 percent from 1985. Recreational visitation for 1987 was 571,964. In 1989, park visits were recorded at 770,114. In 1992, visitation was 885,882 and 913,875 in 1993. In

Photo 4.9. Sunken Forest Nature Walk (Courtesy of Titus Kana, www.FireIsland Vision.com)

Photo 4.10. Sunken Forest (Courtesy of National Park Service)

Photo 4.11. Sunken Forest Platform (Courtesy of National Park Service)

1994, visitation was 810,483 and in 1995 436,067. The decrease between 1994 and 1995 is attributed to a counting change ordered by the NPS (Washington Area Service Office—WASO) and does not truly reflect any change in Seashore visitation. The Fire Island Tourist Bureau reports more than four million visits to the National Seashore annually (this figure includes repeat trips and the developed communities area) and thinks the WASO directive is misleading. In 1996, official visitation was 446,792, in 1997 it was 619,659, in 1998 it was 691,052, and visitation in 2003 was 803,876.[19] See Figure 4.1.

Threatened, Endangered, and Other Species

There are 327 nonmarine wildlife species and over four hundred vegetation species known or expected to be found on Fire Island. According to the 1988 NPS Management Guidelines, parks are required to protect federally listed species. Federally listed species that consistently use park habitat are the Piping Plover (*Charadrius melodus*) and the seabeach amaranth (*Amaranthus pumilis*). The federally listed endangered roseate tern (*Sterna dougalli*) had been sighted in 1993 through 1995 when habitat monitors were trained to look for this shorebird, but has not been observed to breed on Fire Island. A fourth species that has been sighted on Fire Island within the recent past, but that has been restored to nearby areas and could potentially be found on Fire Island, is the northeast beach tiger beetle (*Cicindela dorsalis dorsalis*).[20]

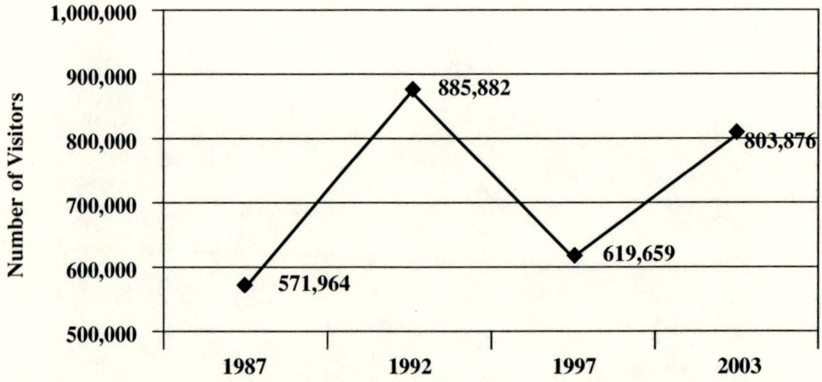

Source: Center for Regional Policy Studies

Figure 4.1. Number of Visitors—1987–2003

Photo 4.12. Wilderness Area Visitor Center (Courtesy National Park Service)

The park is also required to protect New York State listed wildlife, in accordance with the Endangered Species Act. State listed species that use park habitat include the common tern (*Sterna hirundo*), least terns (*Sterna albifons*), and seaside knotweed (*Polygonum glaucum*).

The presence of communities greatly complicates the management of the natural resources on the national seashore. The four thousand homes on the Island are primarily used as vacation homes. Although human occupation is primarily seasonal, impacts associated with human presence, such as vehicular traffic, solid waste production, ground water contamination, and habitat fragmentation, affect the environment of the national seashore. The absence of paved road further complicates these impacts. Though roadlessness seems on its face to promote environmentally sound conditions, it has actually led to a complicated system of permitted off-road vehicle driving. Park staff have frequently reported negative comments from residents and constituents who are likely to be inconvenienced by shoreside habitat management. Three incidents have recently affected endangered species within park boundaries. The first incident occurred in the summer of 1994. An unknown person entered a signed, symbolically fenced, and enclosed Piping Plover nest and stole four Piping Plover eggs about one week before hatching. The second occurred in 1997, when, reportedly, a youth clubbed a Piping Plover to death at Smith Point County Park. A third incident in

August 1997 saw Brookhaven Town employees with a beach driving permit allegedly remove whale bones from a carcass on the beach in front of the Otis G. Pike Wilderness Area.[21]

In 1986, the Piping Plover was listed under the Endangered Species Act as threatened. At around this time, the park began fencing, public education, and annual data collection for the Long Island Colonial Waterbird Survey. An early strategy used in national parks in the former North Atlantic region, including the Seashore, was to permit off-road vehicles to pass through habitat known to be in use by Piping Plovers in "vehicle corridors." But this policy did not stop several Plover chicks from being trapped in ruts and then crushed by passing vehicles in federal areas other than Fire Island. In an effort to reduce the impacts of motor vehicles on shoreside species and their habitats, three "Vehicle Free Areas" and "Pedestrian Escort Zones" were developed and implemented in 1992. This system was also less than effective, partially because of lack of cooperation by off-road vehicle drivers and lack of adequate staff to fully enforce the policy. During this time, vehicles did crush two Piping Plover chicks on Fire Island.[22]

Graduate research on nesting and feeding requirements of these shorebirds started in 1992, and was completed in 1994. From 1992 through 1994 the U.S. Fish and Wildlife Service helped the Seashore

Photo 4.13. Piping Plover Nesting Site (Courtesy of Fire Island National Seashore)

develop the *Environmental Assessment for the Management Plan for Shoreside Species Breeding Habitat.* The plan called for the Seashore to cease using the ineffectual pedestrian escort system, and instead designated the most favorable habitat areas—Sailors Haven and Bellport Beach to Old Inlet—as vehicle free. According to the 1994 plan, if endangered species were found outside of these zones, the park would have taken management actions based on the ecological nature of the situation.[23]

After the first year of implementation, the Suffolk County Police commented to the NPS that they had incurred increased costs and labor contract issues from changing their accustomed patrol routines on Fire Island. In response to these concerns, NPS management decided to explore vehicle travel through Otis G. Pike Wilderness Area. Since this was determined to be illegal, the option was never used.

From 1995–1997, the 1994 plan was modified with yearly consultation with the U.S. Fish and Wildlife Service. But even with the modifications, Seashore management allowed vehicle access, even in areas previously designated as vehicle-free and even during the season when Piping Plovers and terns typically establish territory and perform early breeding activities. The result had been that endangered species breeding and germination productivity has been relatively lower on Fire Island than in areas with similar habitat. In 1998, an *Environmental Assessment for Endangered Species Habitat Management* was completed, recommending that vehicle-free portions of the Seashore be expanded and for longer periods of time. Protective measures that have been adopted since the assessment include limiting or eliminating beach driving in preferred habitat, patrolling for presence of pets (potential predators), and fencing and enclosing Plover and tern nesting areas. Results are measured by monitoring the presence and survival rate of nesting shorebirds and of vegetation.

By 2003, there were twenty breeding pairs of Piping Plovers, fifty-two breeding pairs of Least terns, and two breeding pairs of Roseate terns. In that year, thirty-five Plover hatchlings and one Least tern hatchling successfully fledged. The two Seabeach plant species also increased in number in 2003.[24]

In 1974, Fire Island's deer herd was estimated at fifty; by 1989, it was close to five hundred. Today the Seashore estimates that five hundred to seven hundred deer live on Fire Island. The Seashore, the Biological Resource Division of the United States Geological Survey, the Humane Society of the United States, and the communities of Fire Island have joined together in a cooperative effort to improve the quality of life for both people and wildlife on Fire Island National Seashore. Through this partnership, the Seashore promotes public education about deer, while continuing a research project testing the effectiveness of *deer*

Photo 4.14. Least Tern (Courtesy of U.S. Fish and Wildlife Service)

immunocontraception (birth control) as a means of population control. This research involves darting approximately two hundred does per year with a vaccine that prevents pregnancy. The NPS also has the responsibility to enforce the federal law prohibiting the feeding of wildlife. All wildlife within the boundaries of Fire Island National Seashore are protected under the code of Federal Regulations: 36 CFR 2.2 (2):

All wildlife within the boundaries of Fire Island National Seashore is protected under the code of Federal Regulations: 36 CFR 2.2 (2):

"*Feeding, touching, teasing, frightening or intentionally disturbing of wildlife nesting, breeding, or other activities*" is Prohibited and Subject to Fine under Federal Regulations.

Photo 4.15. Deer in the Ocean (Courtesy Long Island Regional Planning Board)

In addition, the Seashore continues to use and research immunocontraception as a viable means of deer population control. The research involves using the technique of darting female deer with a vaccine called *PZP (Porcine Zone Pellucida)*. The does are darted from a distance; once injected, they are incapable of becoming pregnant. Preliminary deer counts from a concurrent study indicate that in some areas, deer numbers have decreased 50 percent or more from the 1998 level. The park sponsors and facilitates deer management scoping meetings in three separate park areas, working with community members to develop ways of managing the deer herds on Fire Island and the William Floyd Estate.[25]

Additional threatened and endangered species have been identified on Fire Island and the William Floyd Estate, and are carefully considered in the natural resource management of the Seashore. The New York State protected orchid, *Spiranthes vernalis*, has been found in several bog areas on the Island. Though not previously sighted, the federally listed Beach Tiger Beetle, Cincindela dorsalis dorsalis, thrives in habitats like that of Fire Island and Seashore staff suspect it might eventually turn up.

Chapter 5

Seashore Management

The Formative Years: Creation of the Wilderness Area and Its Conflicts, 1964–1978

With the establishment of the Seashore in 1964 "for the purpose of conserving and preserving for the use of future generations certain relatively unspoiled and undeveloped beaches, dunes, and other natural features . . . which possess high values to the nation as examples of unspoiled areas of great natural beauty in close proximity to large concentrations of urban population," there was great expectation and anticipation of what the presence of the only National Park in the State of New York meant to the State in general, and Fire Island in particular. To the residents it meant a total victory. Robert Moses' proposed dune highway was dead, the federal government would implement and pay for erosion control measures, and most important—their homes would be retained. To the NPS it meant the furtherance of national policy to protect the major seashores in the country. To environmental activists and their supporters on Fire Island and Long Island it meant the potential to protect, preserve, and rehabilitate the seashore to a natural state. The views and commitments of the latter two groups were in harmony. The Fire Island residents and/or property owners were more committed to their personal interests in their property than the larger question of maximizing the natural status of Fire Island in perpetuity. Thus, the preamble phrases containing the conflicting goals of preservation and recreational use by an urban population foreshadowed many of the difficulties the Seashore would face in managing the park.

These conflicting goals were embodied in the Seashore's founding documents. For example, the Seashore was to be managed as a recreation area in accordance with approved policies of the NPS. But the enabling law also provided for an eight mile area (popularly known as the "eight mile zone") near Watch Hill to be kept in its natural state. Section 2e specifically authorizes the Secretary of the Interior to acquire property by condemnation "in an approximately eight-mile area from the easterly boundary of . . . Davis Park . . . to the westerly boundary of the Smith Point County Park." Owners of property in this zone were given the option of life tenancy or up to a twenty-five year tenancy, in lieu of vacating the property and selling it to the federal government. This zone and the Sunken Forest area, a typical maritime forest of American Holly, sassafras, juneberry, shadbush, black cherry, and blackgum, are accorded special protection from the incursion of roads and ecologically incompatible uses (section 7a and 7b).[1]

Other conflicts arose from the language contained in its enabling legislation. Under the provisions for a National Seashore, communities and property owners of "scattered improved properties" were expected to retain their present state of development as long as it remained compatible with zoning approved by the Secretary of the Interior. Toward that end, the enabling legislation held that the Secretary may acquire, without the consent of the owners, such lands as are necessary for public access to the beach but may not acquire other lands without consent where compliance with "a duly adopted, valid, zoning ordinance that is satisfactory to the Secretary" had been demonstrated (section 2e).

The Seashore was also charged with acquiring the beach area along the Atlantic Ocean in front of all communities to assure unrestricted public access, but existing communities on the western part of the island were to be exempt from any major acquisitions. Instead, undeveloped lands within existing communities were to come under the purview of local municipal zoning laws approved by the Secretary of Interior. The Secretary was authorized to "issue regulations, which may be amended from time to time, specifying standards that are consistent with the purposes of this Act for zoning ordinances which must meet his approval." Such standards may prohibit certain "new commercial or industrial uses" and promote the "protection and development . . . of land within the national seashore by means of acreage, frontage, and setback requirements" (section 3b). Section 3d states that such regulations must be incorporated into provisions of local zoning ordinances, which will not be approved by the Secretary if they contain "any provision that he considers adverse to the protection and development . . . of the area comprising the national seashore." The enabling legislation provided only the mechanism of condemnation, which requires adequate funding, as a tool of enforcement for zoning standards within the exempt communities.

These provisions give the federal government considerable authority to regulate land use and development on lands within the boundaries of the seashore that have not yet been acquired. Specifically, section 2a of the enabling law authorizes the Secretary of the Interior to acquire by "purchase, donation, transfer, exchange, or otherwise, the lands, waters, and other property, and improvements thereon and any interest therein, within the boundaries of the seashore." The act further provides for land exchanges between the federal government and other interests within the boundaries of the Seashore. Section 10 of the enabling legislation appropriated $16 million for the acquisition of lands. This sum represented a good initial effort, but it was expected that additional monies would need to be appropriated in the future if the full purposes of the law were to be achieved.[2]

The act creating the Seashore preserved zoning authority for four municipalities: the Town of Brookhaven, in which lies most of Fire Island, the Town of Islip, in which the most intensively developed exempt communities on the west end of Fire Island are located, and two villages within the developed portion of the west end, Ocean Beach and Saltaire. It is not difficult to see how the patchwork of land ownership and governmental authority set forth in the enabling legislation makes comprehensive management of the Seashore extremely difficult. While the towns of Islip and Brookhaven and the Fire Island villages of Saltaire and Ocean Beach are responsible for enforcing zoning laws and issuing construction permits within the bounds of the private communities, the NPS is responsible for protecting the island. The legislation also set up a series of management objectives that would lead to conflicts between competing interests with different goals: property owners who want to protect their houses, beachgoers who want access to beaches, environmentalists who want to conserve the wilderness, and local governments that want the revenue from tourism and property taxes.[3]

Overlapping layers of authority also make it possible for interest groups to shop their policy preferences around to gain a sympathetic hearing. Along with the Seashore, the two villages and two towns, the state, and county also had a role in managing the Seashore. The Seashore, for example, was required to achieve a balance between intensive (high density) use and extensive (low density) recreation use. Administration of the intensive use areas of Robert Moses State Park by the State of New York and of Smith Point County Park by Suffolk County was to continue, somewhat vaguely, "in coordination with the NPS." The Seashore was also required to work with the USCE, which is responsible for shore protection, to evolve a protection plan compatible with the Seashore. The role of the USCE in environmental management is stipulated in section 8a: "The authority of the Chief of Engineers, Department of the Army, to undertake or contribute to shore erosion control or beach protection measures on lands within Fire Island National Seashore shall be exercised in accordance with a plan that is mutually acceptable to the Secretary of the Army and the Secretary of the Interior and that is consistent with the purposes of this act."

There were other management objectives outlined for the Seashore that would involve coordination with other jurisdictions. Automobile access was to be restricted to the existing bridgeheads and parking areas at the county and state parks. The Seashore was to make ferry transportation the primary mode of transport between the Long Island mainland and the Seashore and to various points along the Seashore (i.e., lateral ferry transport). This would require land acquisition for terminal headquarters and docking facilities on Long Island's mainland. Foot trails and bicycle paths were to be provided as the major means of travel on Fire Island, as island-length roads were strictly prohibited. This would become a concern within the exempt communities.[4]

The core of most management difficulties for the Seashore concerned the ambiguity over whether the primary function was preservation or recreation. The enabling legislation did not detail recreational activities permitted in the Seashore, but did mandate the availability of fishing, shellfishing, and hunting, except that the latter may be prohibited in designated areas "for the reasons of public safety, administration, or public use and enjoyment." However, the pamphlet entitled "A Report on the Proposed Fire Island National Seashore," dated March 1964, indicated there will be a visitor center, several visitor contact stations, picnic and camping areas, and trails suitable for walking and bicycling.[5]

Despite the lack of details for recreational use, the Seashore drafted an original master plan in1965 that clearly emphasized visitor use of the park. The plan defined the purpose of the Seashore as providing "distinct and unique recreational, interpretive and educational opportunities to residents of the heavily urbanized northeastern United States and elsewhere through the preservation and conservation of the natural scene, through access to it, and through a sensitive blending of recreational and other facilities within the scene."[6] Toward the goal of providing "protection through development" the plan called for additional ferry boat access, facilities for private boat access, bicycle and pedestrian trails, establishment of more "visitor use capacity," additional south shore access facilities, interpretive themes and methods for visitor use, including groups and the mass media. The master plan designated West Beach, Sailors Haven, Talisman, Watch Hill, Whalehouse Point, Pelican Island, Smith Point West, Smith Point East, Pattersquash, and East Lead as major developed areas with either mainline or lateral ferry service access, and one or more of the following facilities: camping, boating, picnicking, emergencies and first aid facilities, stylized camping areas, bathing beaches, bathhouses, interpretive trails, orientation and visitor facilities, maintenance or submaintenance facilities, and park ranger and employee stations. Long Cove and Moriches Inlet were designated for secondary development. Fishing and hunting areas were also envisioned as well as more visitor protection and security in high intensity use areas in existing state and county parks.

The master plan acquisition list provided for the early and sequential development and preservation of the area from Smith Point County Park to the east boundary of Ocean Ridge. The list was developed with an aim toward the early acquisition, protection, and provision of appropriate visitor use of the Sunken Forest-Sailor's Haven area as well as the acquisition and development of the area between Smith Point County Park and Moriches Inlet. After that, sequential acquisition of land for the two remaining major developments, acquisition of low priority areas for development and use, and acquisition of bay islands were to follow.

In October 1964, implementation of the act creating the Seashore began with the opening of an office at the Sayville Motor Lodge in Sayville, New York. The initial staff included Superintendent Henry G. Schmidt, who transferred from Cape Cod National Seashore and Secretary Betty Thompson, who transferred from the Bureau of Sport Fisheries and Wildlife. After a few months, an office was established at 8 West Main Street in Sayville on the second floor over a Woolworth's store.

The first threat to the sanctity of the Fire Island residents was the announcement of the NPS in November 1964 to acquire title to 101 acres as the first acquisition. There were thirty-one houses located within this tract. Fifty-seven of the acres were located in Bayberry Dunes; and forty-four acres of undeveloped high dunes. These properties were all within what eventually became the eight-mile wilderness area. The Regional Director for the northeast region of the NPS, Ronald F. Lee, stated that any construction outside of the existing twenty developed communities, which were to be left alone, would be condemned. Furthermore, Fire Island Seashore Superintendent, Henry G. Schmidt, would be responsible for monitoring construction activities within the Seashore boundaries.[7] Existing homes within the dune area could remain for a period of ten years and the owners retained the right to move their homes to other developed communities that were exempt from condemnation. Negotiations were initiated with owners of unimproved properties on Fire Island and local appraisers were contracted to conduct appraisals. In 1965, additional tract appraisals were received and the first direct payment for seashore lands was accomplished on March 10, 1965 with the purchase of the .183 acre Albert Brower Tract near Whalehouse Point.[8]

Thus, the policy of the NPS was clearly stated and established, a policy consistent with NPS's efforts to achieve as natural a setting as possible for all Seashore properties under their jurisdiction. During the ensuing decade, the original $16 million allocated in 1964 was expended, but the NPS was unsuccessful at expanding acquisitions beyond that.[9] In addition, the Towns of Brookhaven and Islip continued to issue building permits and construction variances that were inconsistent with the objectives of the Seashore. Supporters of the Seashore were frustrated that the NPS was not moving aggressively enough to either acquire lots before homes could be built, or to take strong action against the towns to force compliance with the management objectives of the original plan for the Seashore. In 1972, fourteen property owners filed suit against the federal government on the grounds that the Department of the Interior was not controlling development. The case went against the residents. The decision concluded that the court couldn't interfere with the towns' zoning jurisdiction. It was apparent that if the U.S. Department of Interior couldn't administratively achieve their purpose in controlling development the only remaining strategy would be to buy the undeveloped lots before they could be built on. Resident and environmental supporters of the Seashore put pressure on New York Senator Jacob Javits to come to their assistance, and on August 20, 1974 the Senator introduced a bill to authorize $10 million for additional land purchases in order to forestall further development. However, the issue wasn't solely a matter of money. The first decade of the Seashore's existence was considered by many as the fallow decade. Of the 19,311 acres authorized for inclusion in the Seashore, only 2,733 acres were acquired. About 67 percent of the Seashore lands were acquired before December 31, 1969, and only 736 acres had been acquired since the beginning of 1970. Included in these lands was the 613 acre William Floyd Estate.[10]

In 1975, within the authorized boundaries of the Seashore, 1,048 acres of land were privately owned (5 percent of authorization), 204 acres were owned by local governments (1 percent), and 1,519 acres by Suffolk County (7 percent). The remaining acreage (73 percent) consisted of offshore slopes controlled in whole or in part by the State of New York. Except for that portion of the bay bottom west of Blue Point to the town line owned by the Bluepoint Company, the Town of Brookhaven owned the bay bottom and ten small islands in the easterly portion of the Seashore. Additionally, the town owned a bay to ocean strip on Fire Island at Davis Park known as Leja Beach. Several of the small islands owned by the town were leased to a private hunting club. The Town of Islip owned the bay bottom, fifty-six acre Sexton Island and two bay to ocean strips on Fire Island called Barrett Beach and Atlantique Beach. Barrett Beach, which was later donated to the Seashore, is located within the Town of Brookhaven and, along with Bellport Beach east of Watch Hill, owned by the Village of Bellport, are of particular importance because they transect major federal tracts of natural significance. It is of interest to note that the town boundaries on the bay and mainland do not correspond due to an error in the original Nicoll Patent of 1697. These lands owned by political subdivisions for the most part serve recreational demands of residents and their use is generally compatible with the recreational mission of the Seashore. The distribution of land holdings in the Seashore in June 1977 is summarized in table 5.1.[11]

It was clear that the vast public access to the natural beauty of the Seashore envisioned in the master plan was not achieved. By 1974, only two federal areas on Fire Island had been developed for recreation—Sunken Forest and Watch Hill. Both of these areas are managed by the NPS to provide recreational activity on lands adjacent to natural areas that must be preserved and managed according to restrictions in the enabling legislation (sections 7a and 7b). The Sunken Forest is located between the bay and the high protective secondary dunes and extends from the western Point O'Woods to the western boundary of Fire Island Pines, including Sexton Island. The forest is Fire Island's most unique natural resource and the principal focus of the Seashore's natural-history interpretation. The outstanding natural area between Watch Hill at the eastern most point of Davis Park to Smith Point West constitutes the Seashore's most extensive and diverse natural environment and has been protected legislatively from the incursion of roads and ecologically incompatible developments.

A more serious deficiency was that the NPS hadn't finalized a Draft Master Plan completed in 1973. Without the plan the towns were not obligated to conduct their zoning and development actions based on approval of the Secretary of the Interior, and so the building of homes continued. Mounting frustration over the failure to implement the 1964 objectives was noted in the news media.[12] Criticism of the NPS's lack of progress was becoming more vocal. Former Secretary of the Interior, Stewart L. Udall (1960–1968), who officiated at the official dedication ceremony on Fire Island in 1964, said he "was baffled and distressed over what has happened to his original plans."[13] The ferry terminals in Bay Shore, Patchogue, and Hecksher State Park, and ancillary parking

Table 5.1. Seashore Holdings by Jurisdiction, June 1977

Owner	*Acres*
Public Ownership (state and local)	12,378.19
Private Ownership	1,034.01
Federal Ownership	5,944.35
Total acreage within the Park (land and water)	**19,356.55**

Source: Center for Regional Policy Studies

fields, which were planned as access locations to get to the Seashore, had not been built. Watch Hill, opposite the Village of Patchogue, and Sailor's Haven, opposite West Sayville, were the only two facilities from the original plan that were built.

Senator Jacob Javits and Representative James R. Grover proposed legislation to authorize $10 million for land acquisition, only to have the Department of the Interior oppose the legislation. This inconsistent posture was motivated by presidential budgetary cutbacks which started with President Lyndon B. Johnson and continued with President Richard Nixon. Superintendent of the Seashore from 1971 through 1975, James Godbolt, admitted to "a bit of embarrassment."[14]

The master plan anticipated in 1965 was now scheduled to be finally completed in a scaled down version by the end of 1974. Robert Fresco, an investigative reporter for *Newsday*, traced the ten year history of the Seashore in its failure to meet the original goals. *Newsday* followed the Fresco report two days later with a scathing editorial. The paper said, "Ten years ago the plan for the Seashore looked like a dream. Now we know it was no such thing. It was a hibernation."[15]

In response to the criticism raised by proponents of the Seashore that the NPS was not aggressively pursuing the original goals for land acquisition, planning, and the location of a visitors' center and terminal, Senator Javits introduced a bill in February 1975 to authorize an additional $10 million to enable the NPS to acquire additional property within the dune district zone.[16]

The catalyst which provided the attempt to move forward was not just the frustration felt by the residents and off-island environmentalists, but the expression of their concern in the form of the lawsuit filed by the president of the Fire Island Association.[17] Part of their chagrin was engendered by the seeming inability of the NPS to produce a suitable comprehensive management plan. The master plan of 1965 and the draft plan of 1973 that were prepared in response to the lawsuit were deemed inadequate. The General Management Plan was finally completed in 1975 and submitted for review as required under the National Environmental Policy Act of 1970.

A strong sympathy expressed by the court for the issues raised by the citizens in the lawsuit is found in its decision:

> Efforts by the environmentalists to preserve our natural habitat—in this instance, Fire Island—cannot help but strike a sympathetic chord. Indeed, this is particularly true where, as here, this laudable purpose appears frustrated by a federal statutory scheme, which despite its lofty terms, provides a mere chimera of environmental protection.

The Court concluded its opinion by stating:

> In their justifiable frustration, plaintiffs have sought relief from the courts, but it is clear that only Congress can provide the remedy . . . Nevertheless, precatory though our words must necessarily be, we cannot help but urge those with the power and authority to preserve this gem of an island to halt their procrastination and get on with the urgent business of saving this charming and fragile outpost of nature before the encroachments of haphazard development irrevocably despoils it.[18]

The one clear picture that emerges from the testimony and court response is that the NPS was not dilatory in their commitment to preserving Fire Island. Rather, the problem of the lack of implementation could properly be placed with Congress. Between 1965–1973, the towns granted at least 320 variances to builders and homeowners, yet the Seashore Superintendent was only notified of eighty-six of them, of which he objected to sixty-five as not being consistent with federal zoning regulations. Yet, not one of these properties was condemned. The excuse from the NPS for their lack of enforcement was that they had neither the money nor an approved management plan.[19]

Javits' bill was submitted on April 6, 1975. However, the authorization was only for $2 million.[20] This limited financial support was the major deterrent to the NPS to mount an aggressive acquisition program. Yet, the Seashore Superintendent, Richard W. Marks, did attempt to carry out his understanding of the proper development of the Seashore. Marks was hired after the Draft Master Plan debacle to come up with a more effective General Management Plan, and he successfully shepherded that plan through to completion in 1976 and approval in 1978. The General Management Plan clearly identified the parameters for implementation and included a "Dune Preservation District" as part of its land classification scheme. All dunes not included in the environmental protection/primitive zone were protected from additional damage by inclusion in this district. The Dune Preservation District, as put forth in the plan, is an overlay district on the Seashore and community development districts. Photographed and measured in 1976, the district runs the entire east to west length of the island and is measured from the mean high water line to a line forty feet north of the crest of the primary dune. The district is remeasured every ten years.[21] This Dune Preservation District was consistent with existing federal zoning standards, which in a similar fashion provided for a Dune District that extends from the mean high water line to forty feet landward of the primary natural high dune crest. This Dune District zone prohibits all construction after 1978, with the exception of dune crossings deemed necessary for public access to the beach.

The irreconcilable problem was the conflict between the NPS and those homeowners whose desire to remain was in direct opposition to the objectives of the NPS in preserving the Seashore's natural resources. If there was to be a specially protected wilderness dune area, then it was incumbent upon the NPS to eliminate those uses that were totally incompatible with a "forever wild" natural environment. A reasonable compromise was arrived at between homeowners within the natural wilderness zone who held legitimate title to the land on which their homes were located. Three options were available to this group of owners. If 50 percent or more of their homes were destroyed, the NPS could condemn the property and pay the owner the fair market value, less the amount of insurance paid. Those properties not storm damaged could receive lifetime tenancy before the building would be removed. The third option was to agree to a twenty-five year use of the building. At the conclusion of the lease period, the owner could move the structure to a location excluded from NPS condemnation, or the park service would simply demolish it.

In response to increasing pressure from those who advocated a more dynamic and aggressive program aimed at consummating the reality of a national seashore park, Congressman Thomas Downey, whose congressional district included Fire Island, initiated an effort to establish in law the "Dune District." The idea contained in Downey's legislative proposal was to have the natural zone of the Seashore declared a federal wilderness area. His effort was supported by Chairman of the Subcommittee on National Parks and Insular Affairs, Congressman Philip Burton, who indicated his support for the federal wilderness zone would be limited to the eight-mile reach from Long Cove to the westerly boundary of the Smith Point County Park.[22]

Downey's effort was immediately supported by President of the Fire Island Association, George Biderman, who urged that Congress appropriate more money in order to expand the Secretary of the Interior's ability to condemn more land.[23]

Biderman summed up the frustrations and anticipation of action by stating: "It's about time, after fourteen years, that the Seashore be given the means to get on with its job."[24]

Superintendent Marks was particularly strident in his effort to implement the 1976 General Management Plan by stating the justification for the establishment of the Dune Preservation District.[25] He pointed out that the establishment of the district would prohibit certain uses and activities that endanger the dune system. The protection of the dune network would also provide protection from storm events. He observed that the goal should be to eventually eliminate all residences that create a continual need for public money for relief and protection. The NPS policy should be to establish a "no development" dune district for coastal protection.[26]

Superintendent Marks, in correspondence with the Regional Director of the NPS's North Atlantic Region, endorsed a detailed sixteen page brief with the heading "Why a Dune District." Its overall view was that the request for legislation to create a "no development" Dune District was essential to the NPS's commitment to preserve the natural attributes of the Seashore. He argued against the use of artificial erosion control measures as being less effective than allowing natural processes to be the policy of choice. He cited Paul Godfrey's work on barrier beaches to emphasize his position.

> Certain forms of erosion control, whether designed to control retreat, dune migration, or other forms of instability, can be successful to varying degrees when natural methods are used. In those areas that are currently in a stage of accretion and dune building, fostering the process of planting dune grasses, and some initial sand fencing and fertilization, has been shown to be a very successful management tool. Such natural methods are especially valuable where human disturbance has resulted in massive dune migration problems.[27]

To further bolster his argument in favor of dune protection, he also cited the work of Richard F. Babcock and Richard J. Roddewig in their report to the NPS.

> Construction of buildings on a dune can cause just such wider effects on other parts of its ecosystem though the early effects on the dune itself may be minimal. Activity around a building after it is constructed may destroy vegetation which gives the dune its storm-withstanding stability. The minimal amount of physical dune alteration or destruction necessary to construct a building on a dune may be enough to make the dune more vulnerable to blowouts or storm damage.[28]

In one sense, Mark's treatise bordered on overkill. He also cited the work of eleven other studies, reports, and programs.[29] There could be no doubt or confusion as to his commitment to maximize a policy of letting natural processes guide the management decisions of the Seashore. His strident position on this issue galvanized Fire Island property owners, including those who supported dune preservation, into an opposition force calling for his ouster. The fear was that if Marks prevailed, the USCE would not carry out erosion control mitigation.

The battle lines were now clearly drawn. Property owners within the communities on Fire Island excluded from condemnation were strongly in support of the proposed federal wilderness designation for the Seashore's natural zone. Property owners and non-property owner "squatters" within the district were generally in opposition. This opposition took the form of attacking the Superintendent for his actions. Now that a finalized General Management Plan was in place, the provisions that required local zoning and variance actions taken by the Towns of Islip and Brookhaven to be acceptable to the Secretary of the Interior, raised the specter of the federal government overriding local home rule powers. The issue was raised just prior to the debate over the wilderness designation. Acting Chief of the Land Acquisition Division, Paul X. Colter, notified a homeowner, Vincent Ficara (a resident of New York City) that his effort to build was incompatible with the guidelines for the Seashore. Ficara was put on notice that the NPS was considering condemnation.[30] He countered with an alternate plan which was also rejected. After several legal and political volley's, the Seashore went ahead with condemnation.[31]

Yet the issue was larger than the steadfastness of various Seashore superintendents to enforce zoning regulations. The Town of Islip and the Village of Ocean Beach recognized the imperative of protecting the dune system with specific provisions within their zoning ordinances to prohibit new development in the Dune District. The Town of Islip Zoning Code text read: "The purpose of this restriction is to preserve the ecology of the dunes and grasses and to safeguard life and property of the barrier beach known as 'Fire Island.' "[32] The Suffolk County Department of Planning went even further. They recommended that additional development even in the exempt communities be stopped.

> The primary dune area is a major concern not only in respect to the preservation of Fire Island, but to the protection that the presence of the dune affords the mainland in the event of violent storms. The preservation of the primary dune is possible within the Seashore area. However, much of the problem of erosion exists in the exempt communities. Maximum control over development in the exempt communities could be obtained by purchasing all of the vacant lots located within the primary dune area.[33]

An added input was provided by the USCE. This had a chilling impact on those Fire Island homeowners who had the mistaken assumption that the creation of the Seashore meant automatic support from the USCE in regard to erosion control. Instead, the USCE made clear in their Final Environmental Impact Statement for the Fire Island Inlet to Montauk Point Reformulation Plan that USCE participation was not an automatic occurrence.

> This proposed project is not intended to allow relaxation of zoning standards relating to beach development. In fact, the Corps will not participate in this project until assurances to prevent development on or near the new dunes or other activities that could affect the integrity of the flood protection measures.[34]

Property owners and squatters within the Dune District now felt more threatened than they had been in the fourteen years since the Moses road proposal was killed by the creation of the Seashore; Superintendent Marks was the lightning rod for their antagonism. A mistaken assumption was that if they could force his removal they would then be safe from condemnation. What they didn't acknowledge was that the issue was larger than just one superintendent. Marks was carrying out his obligations faithfully, and whoever would be in charge would be subject to the same responsibilities. In any case, by December 8, 1979, Marks was retired and was succeeded by Acting Superintendent, Richard Weir, followed soon after by William Schenk.

Despite ostensible support for the creation of a "wilderness dune district" there was conflict between Superintendent Marks and Fire Island residents over the issues of homeowner status within the natural area. In a response requested by George Biderman, president of the Fire Island Association, for clarification on zoning issues and condemnation policies, Marks informed him that P.L. 88–587 requires the NPS to address zoning issues in order to properly implement the approved General Management Plan.[35] He mentioned that properties in violation of zoning were not exempt from condemnation. There were a total of thirty properties that the NPS intended to remove. However, Marks' preference was to reach a settlement by negotiation wherever possible. In fact, thirteen properties were resolved through negotiation with only nine others requiring condemnation. It was also his hope that the remaining eight would be resolved through negotiation.

Marks also had a running correspondence with Louis J. Pennachio of P & C Management Co. over a two month period concerning the interpretation of the various statutes in regard to whether or not all existing structures within the boundaries of the Dune District were "grandfathered" and therefore exempt from condemnation. Pennachio's contention was that despite originating legislation and the General

Management Plan, the congressional action in 1978 led to that conclusion. Marks disagreed by indicating that in his opinion, Congress was shortsighted.[36] This was immediately responded to by Pennachio who stressed that in his opinion grandfathering all existing residences was a proper action.[37] Marks countered by stating that the NPS would follow the originating Act of 1964 as revised in 1978. Furthermore, the mandate in P.L. 94–1978 was already reflected in the draft zoning standards proposed on September 18, 1978.[38] This was followed by Pennachio raising a new issue. He challenged the decision taken by Marks to initiate condemnation of houses on substandard sized lots whose owners had filed for variances in order to rebuild.[39] Several more letters followed until Marks' patience ran out. He responded strongly in his last letter to Pennachio stating that he failed to comprehend Pennachio's seeming failure to understand the written word. He stressed that only houses that were outside of the boundaries of the Dune District and met acceptable NPS zoning standards could be rebuilt. Furthermore, the 1964 Act had the stated purpose of the "protection and preservation of Fire Island." The latter amendment of 1978 did not alter or supercede the original legislation. In addition, there was no mention of "grandfathering" for altered properties. He also mentioned that the Brookhaven code of December 17, 1974, prohibited restoration.[40]

The clearly stated purpose of the NPS to remain faithful to the policy of preserving the dune district area did not deter efforts to make end runs around the policy by existing and potential home owners within the nonexempt section of the Seashore. The aforementioned case of Vincent Ficara is illustrative. Ficara, purchased a substandard lot of 0.115 acres (3,800 square feet) and proceeded to build a house in October 1978, in violation of NPS standards. The location of the property was not in any of the exempt communities, or in the developed area of the Seashore. The NPS immediately sought acquisition by condemnation.[41] Ficara then sought assistance from Senator Ted Stevens who sent a letter to the director of the NPS on behalf of Ficara. This was responded to with the clear justification of the agency's condemnation proceeding. It was pointed out to the Senator that P.L. 95–625 passed by Congress on November 10, 1978, does prohibit condemnation within the Dune District—but only if the private property is kept in a natural state. If construction or alteration occurs, the NPS may condemn.[42]

These attempts to thwart the intent to control development and to preserve the natural areas of the Seashore increased the concern of the Seashore's environmental supporters who now rallied in an organized fashion to come to the aid of the NPS by seeking a wilderness zone designation. The Suffolk County Planning Director notified public officials and leading environmentalists to take supportive actions in this regard.

Letters of support and resolutions from the Town Boards of Islip and Brookhaven were sent to Superintendent Marks and members of Congress.[43]

The only opposition came from vested interest property owners on Fire Island. Opposition centered around the Mid-Fire Island Association, an outspoken group representing most of the fifty or so property owners located outside of the seventeen exempt communities. The property owners believed that they had a right to remain because their homes were not doing damage to the environment and that their removal would not significantly enhance the wilderness characteristics of the natural area. Carole Chambers, president of the Mid-Fire Island Association, told the *New York Times* in 1981 that the Seashore was "wasting government money acquiring property they don't need. And they're focusing on single-family homes, which don't do damage, and they're not accomplishing their mandate to control development and prevent new commercial uses."[44] The Mid-Fire Island Association argued for a loophole clause for properties in "single and separate ownership" (substandard sized plots) and opposed any prohibition of renovation and rebuilding.[45] However, the Federal law did not provide any grandfather exemptions for these properties. Their personal target was Superintendent Marks, who to them was clearly the villain.

A second source of opposition was from President of the Fire Island Seashore Advisory Board, Walter G. Reich, who wrote Supervisor Jack Lefkowitz, Brookhaven Town Supervisor, attacking the Town Board for their adoption of a resolution of support for the NPS proposal.[46] He also wrote Senator Javits, thanking him for his efforts to "ward off the overzealousness of local NPS officials," an obvious swipe at Superintendent Marks.[47]

In 1980, Congress created the Otis Pike Fire Island High Dune Wilderness Area encompassing 6.5 miles from Watch Hill on the west to Smith Point County Park on the east. When the Wilderness Area was established, there were approximately twenty private dwellings located within it. These properties were allowed to continue, under provisions authorized by the NPS. Under those agreements, rights of access to the properties were allowed to continue, thereby excluding those areas and their access corridors from inclusion in the Wilderness Area. Additionally, camping was restricted from those areas to ensure that the inholders were provided with the maximum privacy expected under their use and occupancy agreements. Therefore, camping was restricted in the Wilderness Area to two fifteen-person capacity sites, located just east and west of the Old Inlet boardwalk. The last of these occupancy rights terminated in 1992. Structures within the Fire Island Wilderness Area were removed without incident. Camping was thereafter expanded in the

Wilderness Area. Besides these rights holders, potential additional corridors, Cove Nature Trail, Watch Hill horse stable and maintenance storage, Old Inlet, and Smith Point West Handicapped Nature Trail, were incompatible with the Wilderness Area designation. These prohibited uses ceased as use and occupancy agreements have terminated, and these locations were eventually added to the Wilderness Area.[48]

Management Issues

Preservation and the Communities. During the ten years between the adoption of the original master plan and 1975, it became obvious that the Seashore's relationship with the twenty exempt communities had emerged as the primary management difficulty. At the heart of the issue was the divergent views of the Seashore's role held by the communities and the NPS. Many Fire Island residents fought hard for the creation of the Seashore and their inclusion in it to prevent further development, especially of an island-wide roadway favored by Robert Moses. Ideally, from the viewpoint of residents, the Seashore would prevent encroachment on the communities that the state and town could not or would not prevent, while leaving enough autonomy for private property owners to maintain and improve their property. The Seashore, on the other hand, saw its primary objective as preserving Fire Island's natural features while providing the maximum feasible public access. The Seashore's role in preserving the character of the exempt communities was far from clear.

Under the provisions of the act creating the Seashore, the boundaries of the new federal unit were intentionally drawn to include the communities, which prized their isolation. The land within the communities was termed "exempted" from condemnation by the NPS "so long as the appropriate local zoning agency shall have in force and applicable to such property a duly adopted, valid zoning ordinance that is satisfactory to the Secretary" (Sec. 2e). Zoning standards were established by the Seashore on April 2, 1966 that required that towns and villages submit their zoning ordinance to the Secretary for his approval. But in the succeeding ten years, two of the four local jurisdictions never submitted their zoning ordinances for approval by the Secretary of Interior as provided by law. The Town of Islip submitted an amended zoning ordinance to the Secretary in June 1969 and received approval in writing in August 1969. The Village of Saltaire had submitted its zoning ordinance in 1974, but had not yet received approval by 1976, pending completion of the General Management Plan. Within the exempted communities, conversion of buildings to more intense uses, construction of single-family homes on substandard lots, and the subdivision and development of the few remaining sizeable unplatted tracts concerned residents and the Seashore in terms of the threat to the primary dune, the degradation of the Island's resources, and the deterioration of the character of the communities.[49] From 1965 to 1973, about 320 variances were granted by local jurisdictions. The Superintendent of the Seashore was notified of eighty-six variance applications and objected to sixty-five. The Seashore summed up the situation in its environmental impact statement:

> Since the Seashore's authorization in 1964, the relationship between the National Park Service and these [exempted] communities has remained ill-defined. . . . The National Park Service has had little involvement in controlling land use in the Communities. Community zoning regulations have been formulated, and then reviewed by the Federal Government, but variances have been granted by the governing municipalities, leaving the Federal Government with no resource but condemnation—for which no funds have been made available.[50]

On the surface, the four local jurisdictions appeared to have sufficiently strict regulations to control development within the communities and protect Seashore values. But the situation was complicated by the fact that many of the existing lots on Fire Island date from earlier subdivisions of lots which did not meet up-to-date zoning standards for lot size. The Suffolk County Planning Department estimated that of 1,954 lots on Fire Island, 1,759 of them were of substandard size under existing zoning codes. The town and villages felt compelled to issue variances for building on these lots out of fear over the "takings" issue; the legal argument that regulations restricting use of property takes private property without compensation and thus violates the federal and state constitutions. In the first sixteen years after the creation of the Seashore, exempted communities added one thousand homes and housed as many as sixteen thousand persons on a summer weekend.[51]

A report conducted by Richard F. Babcock and Richard J. Roddewig at the behest of the NPS in 1975 found many ambiguities in the enabling legislation that they believed led to these management difficulties.[52] The ambiguity concerned the role of the Seashore in the development and zoning policies of the already existing communities. The report pointed out that while it is easy to see that the existing communities are not "natural features" that the enabling legislation requires the Seashore to preserve, it could be argued that any undeveloped parcels of land within those existing communities, whether they be "postage stamp" lots or unsubdivided tracts, are entitled to Seashore protection. It is also conceivable that lands within existing communities would be of great interest to the Seashore if they in any way influenced the ability of the Seashore to conserve and preserve the surrounding "unspoiled areas of great natural beauty."

The President of the Fire Island Association, George Biderman, was the principle exponent of the view that the preservation of the character of the developed communities, as it existed when the Seashore was established, is within the conservation and preservation purpose of the enabling legislation. The Fire Island Association pointed to the provision in Section 3(b) of the enabling legislation that any zoning standards adopted by the Secretary of the Interior "shall have the object of prohibiting new commercial or industrial uses" as evidence that preservation of community character is a purpose of the Act. That section does not prohibit new residential development and does not explicitly specify maximum densities for future residential growth. In fact, the Babcock report found that nowhere in the legislative history of the Seashore is there an explicit mention of preservation of the character of the communities. "In our opinion," the report found, "there is no basis in either the Act itself or the legislative history for an argument that the preservation in their natural state of scattered, undeveloped lots within developed communities was part of the conservation and preservation purpose [of the Seashore Act]." The report went on to note the "preservation of the character of the communities" is an implied purpose, and means that the NPS must maintain the ongoing low density residential use pattern of development against commercial or high density residential development pressures. "Only if that traditional, low density residential development threatens to harm the unspoiled and undeveloped areas of the Seashore may the Secretary use his powers to stop it," the report stated.[53]

The system of indirect federal regulation of privately owned land within a Seashore's boundaries by a combination of zoning standards and the threat of condemnation to forestall private land use inconsistent with Seashore purposes is referred to as the "Cape Cod Formula." First used in the management of the Cape Cod National Seashore established in 1961, it was incorporated into the Fire Island National Seashore Act with little change. Though it has worked relatively well at Cape Cod, the political circumstances on Fire Island are fundamentally different from that on Cape Cod. The Cape Cod National Seashore boundaries include all the important natural features needing protection, but not the major developed areas. Though privately owned land totals approximately 17 percent of all land within the Cape Cod National Seashore after initial federal purchases were completed, much of that private land was in the form of undeveloped parcels and scattered dwellings. The most developed commercial and residential areas of each town were left outside the National Seashore.[54]

On Fire Island, by contrast, the intensively developed exempted communities fought hard to be included in the Seashore. The regulations adopted pursuant to the Seashore Act (36 C.F.R. Part 28) divide the Seashore into two districts, a "seashore district" and a "developed area district."[55] The standards for development in the developed area district allow for the construction of detached, single-family dwellings and restricts commercial and industrial development when they are inconsistent with the "purposes of the Fire Island establishment act." Paragraph (d) of section 28.3 states: "It is the clear intention of the act authorizing establishment of this Seashore that all land within its boundaries—except certain 'improved property' and property within delineated communities—be acquired by the United States as rapidly as appropriated funds are made available therefore." This specifically excludes unimproved property within the developed communities from the land to be acquired for Seashore purposes. All of which prompted the authors of the Babcock report to conclude that "In our opinion, the history of administrative policy for recreational areas further supports the conclusion that the Park Service in its actions concerning acquisition, conservation and preservation must be guided by a primary duty to the natural, undeveloped areas."[56] For precisely this reason the NPS proposed excising the fourteen westernmost communities from the Seashore in its Draft Master Plan of March 1975.

Draft Master Plan of 1975

The Draft Master Plan, conducted by the NPS's Denver Service Center staff, called for moving the boundary of the Seashore eastward to the western boundary of Point O'Woods to substantially reduce its involvement in matters pertaining to Fire Island's communities. Only six exempted communities—Point O'Woods, Cherry Grove, Fire Island Pines, Water Island, Davis Park, and Ocean Ridge—would have remained within the boundaries of the Seashore. The fourteen western communities were to be excluded. Residential and commercial development in the remaining six communities were to continue to be subject to federal control, and the Seashore would continue to regulate all overland vehicle use for access to these six communities.[57]

The Draft Master Plan sought several new provisions in law, many of them designed to deal with weaknesses in the Seashore's ability to fulfill its management goals. Among them was a provision to permit the Secretary of Interior to seek injunctive relief in the appropriate federal district court against any action inconsistent with the legislative purpose of the National Seashore. While the Seashore possessed condemnation powers, it had been unable to execute this power because the necessary funding was lacking. The power to seek injunctive relief would have permitted the Seashore to enjoin a project at a stage where the cost of subsequent condemnation of improvements would be minimal. The plan included proposals to increase funding for land acquisition, management, and development; the transfer of the remaining six communities from the development district to the seashore district following a storm if more than 75 percent of structures on existing improved properties were destroyed or damaged; an amendment to existing federal standards that would establish a no-development dune district in which any future improvement to private property might subject the property to condemnation by the Secretary.

Along with these proposals to enhance the ability of the Seashore to preserve natural resources was an emphasis on visitor use and access carried forward from the original master plan. The new plan acknowledged the Seashore's responsibility to make the federal lands on Fire Island available to a broader cross section of the public than it had previously served. The plan proposed improvement and expansion of Fire Island's federal beaches to increase their total capacity from the 1975 figure of 7,300 to 17,500 by 1990. Along with the expansion of the existing supervised beaches of Talisman/Barrett Beach, Watch Hill, and Sunken Forest would be added new supervised beaches at Old Inlet Club/Fire Place Beach Club, Whalehouse Point, Lawrence Point, Forge Point, and Great Gun Beach. The total number of visitors using existing and proposed unsupervised federal beaches—notably at Smith Point West (adjacent to the western boundary of Smith Point County Park) and Pattersquash (adjacent to the county park on the east)—was expected to increase from 1,000 in 1975 on a peak summer day to 3,300 in 1990. The plan included proposals to acquire county lands east of Smith Point County Park, as well as acquisition of up to 120 acres on the south shore of mainland Long Island for the purpose of establishing two or three staging areas for the Seashore. Proposed sites included Heckscher State Park and the Republican Club on the west side of Tuthill's Creek in Blue Point. A third site, along the western side of the Forge River near Moriches, would have been required if Suffolk County lands east of Smith Point County Park were acquired, as proposed. Once regular staging areas were established at Blue Point and Heckscher State Park, all regular ferry service to major federal developments west of Smith Point County Park would be provided from these two locations. Lateral ferry service would be provided to link federal developments at Sunken Forest, Talisman/Barrett Beach, and Watch Hill. The plan also sought legislative authorization for an island-long, twenty-two mile hiking and hard-surfaced bicycle trail extending from Point O'Woods to Moriches Inlet.

The Draft Master Plan was presented to the public at three hearings held in Patchogue, New York City, and at Watch Hill in 1975. The plan met with strenuous opposition on several, sometimes contradictory, fronts. For the most part, Fire Island residents viewed the Seashore as their only recourse to the dictates of often unresponsive town governments. The peculiar demographics of Fire Island is important to understanding why. Geographically, Fire Island constitutes a tiny fraction of Islip and Brookhaven towns and is separated from the mainland by the Great South Bay. Though there were approximately twenty-seven hundred dwellings on Fire Island when the Seashore was created in 1964, only one hundred were permanent, year-round residences. The approximately 120 families living year-round on Fire Island constituted a politically insignificant portion of the two towns combined population of over five hundred thousand residents. Most owners of summer homes on Fire Island were a distinct minority in Suffolk County. Islip and Brookhaven include more than 60 percent of the entire population of Suffolk County, yet Suffolk County residents constituted only 6.5 percent of the owners of summer homes. Somewhat justifiably, then, Islip and Brookhaven town boards traditionally regarded Fire Island as an isolated enclave of wealthy Manhattanites whose governmental and land use problems ranked low in political priority.[58]

Residents therefore greeted the proposal to remove all but six communities from the Seashore with great derision. George Biderman blasted the proposal as "an open invitation to real estate speculators to move in." To the Fire Island Association the Draft Master Plan was "a political proposal. It represents the administration's retreat from the 'parks to the people' concept and says that the National Park Service is unwilling to cope with the pressures of private real estate interests." Suffolk County Executive John V. N. Klein, called the westward boundary change "a bail-out" by the park service. "They're simply taking the easiest way out of a difficult situation," Klein told the *Long Island Press*.[59]

Further evidence of the difficult position the Seashore was in comes from community opposition to other parts of the Draft Master Plan. While residents attacked the plan for "caving in" to developers on the western end of the Seashore, some of them also attacked the plans strongest conservation proposal—the proposed Dune Preservation District. The Dune District was to be a 100-foot-wide zone that would run the length of the Seashore and extend from the seaward foot of the primary dune back toward the center of the island. Under the proposal, no future construction would be permitted and the existing houses would be declared nonconforming. Owners would be allowed to remain in possession for up to twenty-five years but would not be able to transfer the property to heirs. Eventual condemnation and acquisition of the houses would be the responsibility of the municipalities on Fire Island—the villages of Ocean Beach and Seaview and the towns of Islip and Brookhaven—rather than the NPS.

But at an August 1976 meeting four hundred Fire Island residents railed against the plan for the Dune District. Speakers told of assiduous efforts to prevent erosion of the dune by owners of houses on it, depicting themselves as ecologically effective caretakers of their own property, and denouncing condemnation and acquisition efforts. Charles Collingwood, a television newscaster and a dune homeowner, said, "The proposal for expropriation is bureaucratic and arbitrary. Where it can be shown that a house is harming the dune, it should be removed. But the bull-in-a-china shop condemnation of all dune structures is unnecessary and besides, it would be too costly."[60] Somewhat paradoxically, though, while critical of the Seashore's attempt to form a protective dune district, residents were supportive of the plan's proposal to install sand bypass systems at the Moriches and Shinnecock Inlets, which would replenish beaches at the west end of the island, where the developed communities were located.[61]

To a large extent, many of the criticisms of the Draft Master Plan reflected a split over the perceived purpose of the Seashore and the constituencies it was designed to serve. In eliminating the western portion of the exempted communities from the Seashore, federal officials believed they were improving the NPS's ability to provide public recreation by freeing management to concentrate in areas of greatest public involvement. The Seashore, in other words, had not abandoned its goal of becoming a high-use park managed by NPS guidelines for recreation areas. Other objections to the Draft Master Plan seem to bear this out. The Fire Island Association loudly protested the Draft Master Plan's proposal to turn the Fire Island Lighthouse over to New York State out of fear that it would become a high-use recreation zone. "To relinquish this 90-acre tract to Robert Moses State Park for high-density recreation would be to obliterate its unusual natural values," the Association said. For the same reason, the Association objected to the plan for a bikeway-walkway spanning the length of the island. Such a trail would "bring an increase in harassment of residents, conflict between residents and visitors and increase police problems," the Association complained.[62]

To the residents of the Fire Island Association, the Seashore was viewed not only as a bulwark against development in exempt communities, but it was also viewed as a potential threat because of the Seashore's objective of providing public access to the beach. To a certain degree, residents wanted the Seashore to become a babysitter and policeman for their own private interests, instead of a public park. The residents who make up the Fire Island Association seemed perfectly willing to use the Seashore as a foil against further development in exempted communities, something the Seashore would rather have left to the towns. The Fire Island Association's opposition to the Seashore as a heavy use park was echoed by environmental groups, who also criticized the Draft Master Plan. A comprehensive sixty-four page report put together by the Fire Island Natural Area Committee recommended that the Seashore eliminate its attempt to develop lands east of the Smith Point County Park for visitor use. They also recommended purchasing larger swaths of land between west end developed communities and to keep camping facilities elsewhere on the island to a bare minimum. The Committee also recommended that the eight-mile zone be designated as a "wilderness" to permanently ensure its protection. Most importantly, the Committee criticized the NPS for encouraging formal visitor use of park facilities as a means of gaining larger appropriations for itself. "The park service thus seems to function as if it were a commercial enterprise, marketing a product—recreation," the Committee claimed.[63]

As a result of this opposition the NPS proceeded to make several major modifications to the 1975 Draft Master Plan. The first step was to replace the superintendent of the Seashore, a position that has an interesting history of its own. The first Superintendent of the Seashore was Henry G. Schmidt. In January 1964, the NPS properties located in the New York metropolitan area were grouped into one administrative unit known as the New York City Group (NYC Group). This NYC Group, made up originally of the Statue of Liberty, Federal Hall, Castle Clinton, the General Grant and Hamilton Grange National Memorials, and the Theodore Roosevelt Birthplace and Sagamore Hill National Historic Sites, was headed by John A. Townsley, who appointed management assistants to administer on-site at the various federal properties. In 1967, the Seashore was added to the NYC Group, changing the Group's name to the Fire Island National Seashore and New York City NPS Group (NYCNPS Group). At that point Schmidt replaced Townsley as superintendent of the NYCNPS Group and Thomas Norris became superintendent of the Seashore. In the early 1970s, this administrative body became so cumbersome that Seashore split off to become its own administrative unit, and the NYCNPS became known, briefly, as the New York District Office. In 1971, Tom Norris was succeeded by Jim Godbolt, who was responsible for the Draft Master Plan. He was replaced in August 1975 by Richard Marks, who in turn would be followed by five superintendents (see appendix A), but it is Marks who is credited with making several important changes to the planning process.[64]

In the first place, the Seashore adopted the concept of a general management plan, designed to be more responsive to current planning needs and to provide more flexibility for public involvement than the master plan concept, which it replaced. Consequently, the Draft General Management Plan reduced its emphasis on increasing visitor use and put greater emphasis on preservation.

The Draft General Management Plan organized the Seashore into seven activity areas oriented toward a variety of recreational, interpretive, and educational uses, but reduced projected daily visitors at existing and proposed federal activity areas from 17,500 by 1990 to 8,470. "The Park Service recognizes its responsibility to make the federal lands on Fire Island available to a cross section of the public while placing primary emphasis on resource preservation. Therefore, day-use recreational beach areas rather than additional marina development will be emphasized," the plan stated.[65]

The seven activity areas included the Fire Island Lighthouse Tract (not yet in federal hands), the Sunken Forest, Talisman, Watch Hill, Old Inlet, Smith Point West, and the William Floyd Estate. While proposals for behind-the-dune hiking trails at some locations remained, the proposal for a bike trail the entire length of the island was deleted.

Administrative policies for recreation areas of the National Park System require the use of a land classification system proposed by the Outdoor Recreation Resources Review Commission and recommended for use on Federal lands by the Bureau of Outdoor Recreation. This system places all lands within a park in six classes, defined according to characteristics of the land and desirable uses. In its 1978 General Management Plan, the Seashore utilized a modified version of this land classification system to make it more specific, and therefore, more useful to the Seashore. The land classification in the General Management Plan took into account the suitability of certain areas for specific uses. Four use zones were identified for the national seashore: natural, historic, general outdoor recreation, and special-use. The listing below gives a summary of the land classification system

Class I—Natural Zone

In this zone, natural resources and processes remain largely unaltered by human activity except for approved developments essential for management, use, and appreciation of the park.

Sub-Class 1A—Environmental Protection/Primitive Zone. Lands or waters possessing particular value as wildlife habitat and/or research areas with Outstanding physical and biological characteristics.

Sub-Class 1B—Outstanding Natural Feature Zone. Geological and biological features possessing intrinsic value or uniqueness.

Sub-Class 1C—Natural Environment Zone. Lands that require protection of the natural environment but can accommodate environmentally compatible activities.

Class II—Historic Zone

This category includes areas managed to preserve and interpret significant historic structures as well as lands containing resources listed on or eligible for the National Register of Historic Places. Authenticity and integrity of the historic scene should be maintained with compatible uses of all historic structures. First order structures are normally restored and preserved. Historic structures of lesser importance can be used for a variety of compatible purposes.

Class III—Park Development Zone

Identified as the General Outdoor Recreation Zone in the General Management Plan, this zone includes lands and waters where visitor use has altered or may moderately alter the natural environment. Facilities on lands within this zone serve the needs of park managers in providing medium-density recreational opportunities. Facilities and structures would not dominate the natural environment and should be architecturally compatible with the resource.

Class IV—Special Use Zones

Sub-Class 4A—Community Development Zone. This sub-zone includes private lands within the seventeen communities located within the boundary of the Seashore. Federal control within the communities is limited by existing legislation to certain zoning standards. Otherwise, the communities are exempted from direct federal control.

Sub-Class 4B—Non-Acquired Park Development Zone. Lands within this sub-zone are limited to major large-scale facilities, public bathhouses, and parking lots located within Smith Point County Park, Suffolk County. Being within Special-Use Zone, this applies only to non-NPS owner areas, as compared with NPS lands within the Class III Park Development Zone.

Sub-Class 4C—Dune Preservation District. (This zone is now considered part of the Natural Zone, but the General Management Plan includes it in Special Use Zone). The dune system of Fire Island National Seashore is vital to the protection and perpetuation of the resource. Certain segments of the dune system within the Seashore have been severely damaged by human activities. All dunes not included in the environmental protection/primitive zone will be protected from additional damage by inclusion in this district. The Dune District is an overlay district on the Seashore and community development districts. Photographed and measured in 1976, the district runs the entire east to west length of the island and is measured from the mean high water line to a line forty feet north of the crest of the primary dune. The District is to be measured every ten years.[66]

The Draft General Management Plan also eliminated the proposal to remove western exempted communities from the Seashore boundaries, as well as the proposals to hand over the Fire Island Lighthouse tract to Robert Moses State Park and to acquire land east of Smith Point County Park. The ninety acre United States Coast Guard Lighthouse tract was to be acquired, with the Fire Island Lighthouse made an historical and environmental center. Visitation to Talisman was to remain limited until federal facilities could be constructed there. Sunken Forest and Watch Hill were to remain the primary federal destinations for visitors. Private vehicle access to the Seashore (except by property owners with permits) was to be limited to parking areas at Robert Moses State Park on the west end of the island and Smith Point County Park at the east end. Visitor access to Seashore facilities at the Sunken Forest, Talisman, and Watch Hill were to be by means of ferry, small boat, or foot. Visitors to the Lighthouse area and Smith Point West were to enter those areas on foot from adjacent state and county parking areas. Major ferry routes to federal areas were to originate from a proposed new ferry terminal on the Patchogue River and from existing private ferry operations in Sayville. The Seashore committed itself to study the feasibility of ferry terminal development at Heckscher State Park to replace existing ferry service from Sayville.[67]

The revised plan reiterated that the prohibition of additional private land development within the twenty exempted communities was not intended by the Seashore's enabling legislation. However, the revised plan did propose the promulgation of a model zoning ordinance that would set standards for single-family residential development, prohibit conversion to multifamily dwellings, and evaluate commercial and industrial properties for possible down zoning.[68]

Along with acquisition of the ninety acre U.S. Coast Guard Fire Island Lighthouse tract through surplus property procedures, the revised plan recommended acquiring 49.15 acres in seventy-nine parcels within the present Seashore boundary, as well as a bay-to-ocean strip known as tract 2203, consisting of 8.5 acres. The revised plan also called for the acquisition of ten acres of land at the head of the Patchogue River for the purpose of establishing a new headquarters. This amended the old plan's proposal to establish a primary ferry terminal/headquarters site at the Republican Club in Blue Point, which drew heavy opposition from the mostly residential neighborhood of Blue Point. The revised plan established a wetlands district to protect fragile wetlands on the bayside of the island, and retained the "high dune preservation district" in the eight-mile zone. The dune district proposal called for an end to new construction on or near the primary sand dunes and for the forty-eight existing houses on these structures to be torn down. The eight-mile long dune preservation district would be classified as a primitive area, with minimum facilities provided. The idea of seeking injunctive relief against nonconforming construction was retained, and a provision was added to acquire lands within exempted communities if major storm activity destroys 90 percent or more of all structures within a community. The revised plan also discontinued plans for active use of the small federal bay-to-ocean strips between the western communities, and changed the land classification system to further limit permitted uses and facilities. The General Management Plan was intended to guide management, development, interpretation, and use of the resources of the National Park System for a period of approximately ten years, until about 1987.

The revised Draft General Management Plan was received with much greater enthusiasm by the public at hearings held in Patchogue, Bay Shore, and Ocean Beach in 1976. Suffolk Legislator John Foley of Blue Point praised the revised plan and the new Seashore administration for the "very basic change from heavy federal development to resource preservation." Thomas S. Zawyrucha, executive director of the Marine Environmental Council of Long Island, praised the decision to keep the Seashore land east of Smith Point in its "primitive" state and praised the plans for protection of the lighthouse. "The dune is the essential factor in preserving Fire Island as a barrier beach," said Zawyrucha. Head of the Fire Island Natural Area Committee, Joe Zysman, which had led the fight to block development of the Seashore's eastern sector, said, "we wish to commend the planning team and the NPS for their responsiveness to public concern that has resulted in the extensive revisions we see in this draft." Representing the Natural Resources Defense Council, Frances Beinecke, said the draft "finally represents a management plan which recognizes the existing uses of the Seashore and the primary interests of its traditional users."[69]

Opposition to the plan still remained. Residents of developed communities still opposed the dune preservation district. Sportsmen and workers on Fire Island also opposed the Seashore's plan to limit vehicle use on Fire Island "to that which is absolutely necessary." In an accompanying set of regulations, implemented on January 1, 1978, the Seashore sought to limit the residents of these communities to one round trip a day by beach buggy to the mainland of Long Island via the Robert Moses Causeway.[70] During the summer, as in past years, no travel at all was to be allowed.[71] The groundwork for beach buggy regulation was laid in May 1965 when numerous conferences were held with representatives of the Great South Beach and Long Island Buggy Associations to discuss regulations of oversand vehicles on Fire Island. A close review of the beach buggy regulations used during the 1967 season resulted in extensive revisions in 1968 designed to eliminate inequalities and strengthen the ordinance. During the winter of 1967–1968 a joint Seashore, Town and Village Committee chaired by Suffolk County Commissioner of Parks Charles R. Dominy laid the groundwork for uniform revisions of local beach buggy regulations. In 1971 and 1974 these regulations were further amended, and the current regulations appear in Section 7.20, Title 36, of the Code of Federal Regulations. The regulations prohibit the operation of vehicles across Seashore lands except under permit issued by the superintendent and restrict their operation on a seasonal, daily, and hourly basis. Rules and routes of travel are laid down and penalties are imposed for violation of the regulations including, but not limited to, suspension or revocation of permits.

But overall the new Draft General Management Plan, and Superintendent Marks who directed it, was greeted with high accolades, leading to the adoption of the General Management Plan in 1978. Marks arrived at the Seashore in August 1975 and was credited with using extraordinary sensitivity and diplomacy in drawing the narrow line between natural preservation and public access. Marks believed he had a better understanding of what the Seashore's mission was than previous administrations, which he described as preserving "Fire Island as best as can be done" while providing ways for some of the region's mass humanity to see the Seashore, but to visit it gently. "If you want a mass beach type of thing," Marks told the *New York Times*, "you either go to Smith Point or you go to Jones Beach. The purpose of the National Seashore is to give a person an opportunity to walk away from people."[72]

Land Use

The enabling legislation for the Seashore outlines the acquisition and regulation as the two land protection techniques for use at Fire Island. The NPS is authorized to accept property by donation, transfer, and exchange; acquire undeveloped lands within the "Seashore District"; and prepare standards by which the adequacy of local zoning ordinances will be evaluated. Except in the eight-mile zone, the Seashore is directed not to acquire "improved properties" that are maintained in accordance with zoning ordinances approved by the Secretary of Interior.[73]

As previously noted, most of the original $16 million authorized for land acquisition was spent by 1969. A 1965 legislative amendment authorized the Secretary to accept donations of lands and structures at the William Floyd Estate for inclusion as a detached unit of the Seashore. By 1977, the Seashore had acquired in fee, or easement, 5,944 acres of land within the Seashore's boundaries. While the dollar ceiling for land acquisition was increased by Congress in 1976 to $18 million and to $23 million in 1978, only small additional amounts were actually appropriated and spent. The flow of funds from the federal Land and Water Conservation Fund (LWCF) dwindled to only a trickle in the late 1970s. Ironically, it was LWCF funding that originally provided the resources for the establishment of national seashores in the park system.[74]

In 1978, Congress enlarged the Seashore's boundary to include New York State and U.S. Coast Guard properties in the vicinity of the Fire Island Lighthouse and twelve acres of land on the Patchogue River in Patchogue for the new Seashore headquarters/terminal.[75] The law added island and marshlands in the Great South Bay, Bellport Bay, and Moriches Bay (Sexton Island, West Island, Hollins Island, Ridge Island, Pelican Island, Pattersquash Island, and Reeves Island) to the Seashore. This same legislation also authorized federal condemnation of properties within a "Dune District" under specified conditions. Legislation enacted by Congress in 1980 designated an approximately six-and-one-half-mile length of the barrier island as wilderness, to be managed consistent with National Wilderness System policies.

The Squatters Controversy

A bitter dispute arose between the Seashore superintendent acting on behalf of the Seashore and twenty-one individuals and families who had built unauthorized shacks and cabins within the area that was outside the established communities that were safe from condemnation. These structures were primitive, lacking electricity, inside plumbing, winterization, and so forth. After being notified that these structures were situated on land belonging to the United States government and that they would be evicted, they chose to challenge the NPS in court.

Twenty-one of the squatter families took action as plaintiffs against Seashore Superintendent Marks and Secretary of the Interior Cecil B. Andrus in the United States District Court for the Eastern District of New York on August 2, 1978. They sought injunctive relief to set aside the previous stipulations that they had agreed to years earlier. They claimed that they were entitled to the same privileges that other residents who actually held title to the land their buildings were on. Their argument was that they owned the homes and that the Seashore Act did not specifically require that the owners had to also own the property on which the buildings were located. The Cape Cod Seashore legislation did clearly state the requirement of land ownership. Their argument was specious on several points. Since the argument of adverse possession didn't apply to these squatters, and the land did belong to the United States government, it was implicit that home ownership meant legal title to the land as well as the building. Furthermore, comparison with Cape Cod was not analogous or applicable since the Seashore did include established residential communities within the legal boundary of the Seashore. Conversely, the Cape Cod National Seashore explicitly excluded existing communities, and basically incorporated undeveloped property. They based their case largely on correspondence sent by John A. Carver, Jr., Assistant Secretary of the Interior to Representative John Lindsay of New York City, in which is stated, "... the Park Service would not be inclined to disturb authorized occupancy until the land is actually required for Seashore purposes. Even at this point, however, every effort would be made to reach reasonable and amiable agreement as to the time when actual possession is to be transferred."[76] A key word in Carver's comment was "authorized." Squatters were not considered authorized occupants. Nevertheless, the NPS did extend continued existence for a period of ten years which the squatters had agreed to. The injunction was denied.

On September 13, 1978, Judge John R. Bartels of the Federal District Court ruled that the NPS was within its rights to evict these "squatters." One of them, Irene E. Sorse, wrote to Senator Javits asking him to introduce legislation in Congress to enable them to retain their occupancy.[77] One of Javits' staff aide wrote to Senator Javits asking him to decline to intervene.[78] She pointed out that in addition to the "squatters" there were another twenty-five "legal" families whose dwellings were on lots owned by them, in contrast with these who had no legal claims to the land. The property owners could remain as the result of negotiations in 1968 with the NPS for a period of twenty-five years, until 1992. Her memo also noted that the organized groups were not in support of the squatters, and they also did not agree with the Senator when he stated his opposition to grant additional condemnation powers to the NPS This recommendation was countermanded by Barbara's boss, Beverly Charles, advising Javits to send her draft letter to Secretary of the Interior Andrus, stating Javits' support to treat the squatters with the same benefit negotiated with the legal property owners.[79] She obviously was in direct contact with Daniel L. Alterman, Esq., the attorney for the squatters, since he sent a letter to Beverly on the same date, thanking her for her support. He also reminded her that there was a deadline of November 15, 1978, and that the Senator should act swiftly.[80] On the same date, Javits wrote to Alterman and cited the letter that he sent to Secretary Andrus requesting that the Secretary issue a stay order to prevent the NPS from action until Congress "may or may not think that legislation is warranted." If the facts indicate that the squatters should be supported, the Senator would then introduce a bill to that effect.[81]

The correspondence seemed to indicate a win situation for the Senator. By sending the letter he was catering to the squatters and their legal counsel. Yet the iffy nature of his comments was not an outright declaration of his total opposition to the efforts of the NPS to acquire the land. This "wink-wink" approach recognized the Secretary's authority to proceed and was undoubtedly interpreted as such, since the NPS was proceeding to destroy the squatters shacks.[82]

One of the shack owners, Mr. Edward Rant, whose dwelling had no utilities, claimed that although the land had been purchased a decade earlier by the federal government, he thought the land was "open land" and, therefore, erected his building without title or permission. However, since 1967 the NPS had been easing out the residents within the District between the community of Davis Park and Smith Point County Park. Some properties were bought outright. Some homeowners had the option of remaining until 1992. The squatters were granted permission to remain until Labor Day 1977 before they were required to vacate. Mr. Rant actually had signed the agreement to vacate earlier, but now that the deadline had passed, was suing the NPS The Court did grant a stay from their December 1, 1978 deadline in order to afford Senator Javits the time to request the Congress to intervene.

Secretary Cecil D. Andrus responded to Senator Javits in a letter dated November 15, 1978, in which he rejected Javits request to delay action on the twenty-one squatters' cottages.[83] This correspondence was followed by a letter from Acting Secretary, Robert L. Herbst to Senator Javits in which he stated three reasons for the NPS to proceed with the eviction.[84]

1. The condemnation powers granted to the NPS in the eight-mile Dune District zone demonstrated the Congressional intent to restore the area to natural conditions (16U.S.C.459e, et seq.).

2. The squatters are not entitled to the same considerations as the land owners.

3. The policy being pursued at Fire Island is the same policy for all Seashore parks.

The squatters were sued in Federal District Court to uphold the NPS order of September 13, 1978 to surrender their dwellings located on federally owned land Judge John R. Bartels' decision was rendered on December 1, 1978. He stated that these were not "homes" but recreational cottages and some were not even occupied seasonally, but rented out. The squatters wanted compensation for land they didn't even own. They had been granted continued use for a period of ten years. Justice Bartels' final comment displayed the Judge's disdain for the arguments posed by the squatters:

> We find no reasons, in equity and good conscience, why respondents should be permitted to remain in the Fire Island National Seashore beyond the time agreed on. It appears to us that respondents who have spent their vacations for free on another's property for decades, have been dealt with more than fairly and in strict accordance with the law by the Park Service. Persons who make promises and fail to keep their word have no standing in equity before this court.[85]

In the verbatim record of the trial before Judge Bartels, the attorney for the squatters (defendants) was Mr. Kenneth Kimmerling, Esq., of the New York City law firm of Stolar, Alterman & Gulielmetti, Esq. Edward R. Kormer, Esq., the United States Attorney for the Eastern District of New York, sent his Assistant U.S. Attorney, Cyril Hymen, Esq., to represent the NPS, who were the plaintiffs. A reading of the record covering the lengthy exchanges between the Judge and Mr. Kimmerling revealed the increasing impatience of the Judge, who expressed his belief several times that the attorney was merely trying to delay the inevitable and without any legal substantiation for his position. In response to Kimmerling's request for an approval to appeal Bartel's ruling, the Judge said: "I don't think there is any merit whatsoever to your appeal. And, I think it's another ploy to delay the full surrender which both the law and equity require you to make . . . with a requirement that you request an immediate accelerated appeal."[86]

An examination of the chronology of events between the NPS and the squatters clearly identifies the chagrin of both the NPS and the Judge in trying to arrive at a conclusion to the issue. It indicates that the NPS and the Judge showed great forbearance in dealing with the squatters. First of all, the NPS gave the squatters two extensions to the order to vacate. Finally, a registered notice was sent to each squatter on August 25, 1977, informing them of the expiration of permission to remain. The letter announced that the NPS would hold a meeting on September 8, 1977 wherein the NPS would stress that the removal of the squatters was the course the NPS would follow. In response for reconsideration from the squatters, the NPS said reconsideration would occur—but—the squatters must take steps to remove their property. On December 28, 1977, NPS announced that it had completed it's reconsideration, and they once again rejected the squatters' requests. They were then all notified by certified mail that they must vacate by February 15, 1978. At the request of the squatters, another meeting was convened on January 24, 1978, at which the NPS held firm. Instead of complying, the lawyer for the squatters once again filed another petition for reconsideration, which was denied by the NPS on May 8, 1978. The squatters once again received letters to remove their property by May 29, 1978. When their counsel said another ninety day notice was required, the government lost patience and instituted the court action.

Despite the litigation, the squatters were allowed by the NPS to remain for the summer season until September 1978.

It must be noted that a decade earlier, in 1968, the respondents did file settlement agreements with the federal government which stipulated their disclaims to the property. The filing included consent for the taking. The legal firm of Hollman, Wright, Glutt and Brown, who filed the consents, was now a decade later challenging their own agreements.

The expedited appeal was heard by the United States Court of Appeals for the Second Circuit and a decision rendered on December 15, 1978.[87] Assistant United States Attorney, J. Christopher Jensen, argued that Judge Bartel's decision be upheld and that the writ of injunction pending appeal be dissolved. He argued that the appellants' case was not timely. Nor were these seasonal buildings the defendants' "homes" but merely recreational cottages, and in fact a number of these cottages were not even occupied seasonally by the defendants; they were rented out during the summer season.

As the appeals case was being heard, Senator Javits' staff advisor, Beverly Charles, sent a memo to the Senator with an enclosure of the newspaper accounts of the case. She also indicated that the attorney for the squatters, Ken Kimmerling, wanted the Senator to issue a press release in support of his clients. She advised, "whatever your decision on this issue, you will still stand as the hero of the squatters if the media's reaction is an accurate barometer."[88] Kimmerling had called Charles on December 18, when he presented his arguments to the court and expressed his opinion that an immediate release to the press by the Senator might influence the Court.

The same day Charles sent a second memo to the Senator, stating that she received a second call from Kimmerling informing her that the court was going to affirm Bartels' decision and therefore the Senator could disregard the request for a press release. Demolition of the squatters shacks would resume on December 18, 1978.

Seashore Zoning

In the Act that established the Seashore, the Secretary of Interior was directed to issue zoning regulations to bring development in Fire Island into conformity with the goals of the Act itself. Federal zoning standards for the Seashore are contained in Section 28 of CFR Title 36. The standards, which were amended in 1980 and 1991, set guidelines for local zoning authorities to use in developing local codes that conform to the federal regulations. The federal zoning standards divided Fire Island into three land use districts. The Community Development District, comprised of the seventeen exempt communities on the Island's west end, generally permits construction or expansion of existing residential units, religious institutions, schools and commercial units in existence before 1964. The Seashore District, comprised of all portions of the lands and waters within the boundary of the Seashore which are not included in the Community Development District with the exception of the headquarters facilities at Patchogue and the William Floyd Estate at Mastic, permits alterations of existing improved property, but prohibits new construction. The Dune District, which extends from the mean high water line to forty feet landward of the primary natural high dune crest, prohibits all construction after 1978, with the exception of dune crossings deemed necessary for public access to the beach.

In 1985, the Secretary of Interior approved compliance with federal zoning standards for the four zoning authorities within the Seashore. Some of the municipal zoning codes were approved with exception, in that their noncompliance was not viewed as significantly detrimental to the natural resources of the Seashore and, therefore, should not be disapproved by the Secretary.

The late 1970s and the early 1980s were a critical time for establishing federal authority on the Seashore in terms of zoning. The issue had been fiercely contested by property owners outside of the seventeen exempted developed communities during the debate over designation of the Wilderness Area. But condemnation also became a tool with which zoning regulations were enforced in the developed communities. In 1980, the Seashore amended the federal zoning standards to prohibit any increases in lot coverage in the undeveloped portion of the Seashore beyond what it was in 1963. The one condemnation case that was crucial to the Seashore's efforts to make these new federal zoning standards stick was that of homeowner Fredric Rose, a beer wholesaler from Patchogue who bought a vacant double lot in Fire Island Pines in the 1970s for $100,000. Rose sought a variance because he lacked 400 square feet of the 7,500 he needed to build a house overlooking the Great South Bay. The Seashore threatened to condemn Rose's property if he obtained a variance. In June 1981, Judge Eugene H. Nickerson of Federal District Court in Brooklyn upheld the Seashore's power to take Rose's land and ruled that the condemnation could proceed. The Seashore, originally under Superintendent Marks, then under his successor Joel Pickelner, condemned Rose's property three different times. Rose ultimately ended up buying back the condemned lots from the NPS and building on them in conformity with the zoning standards. But the message to property owners about the seriousness with which the Seashore took its zoning standards was clear.[89] "Unquestionably," says Donald Weir, former acting superintendent who served under both Marks and Pickelner, "the Rose condemnations were central to the Seashore's effort to consolidate its authority over zoning on Fire Island."[90]

Nevertheless, pressure mounted on the Seashore to revise its standards. In 1980, U.S. Senator Daniel Patrick Moynihan and former Senator Jacob K. Javits requested that the United States General Accounting Office conduct a study of the Seashore's enforcement of zoning standards. The report, released in June 1981, found that the NPS was "unnecessarily acquiring private lands on Fire Island" and that "the Park Service should revise its zoning standards to comply more closely with Congress's intent and should sell back to private citizens lands it has acquired but does not need."[91] As a result of this kind of pressure, the Seashore revised federal zoning standards in 1991.[92]

The 1991 amendments relaxed the federal zoning standards somewhat. The revised standards removed setback requirements in favor of focusing on the 35 percent lot coverage standard. Lot occupancy of all privately held improved property in the Seashore District was limited to 35 percent of the square footage of a lot that is less than 7,500 square feet, and to 2,625 square feet for a lot 7,500 square feet or greater. Lot occupancy is calculated to include all buildings and accessory structures on the property and any extension of the upper floors beyond the developed area of the ground.[93] This change was instituted with an eye toward getting out of the "metes and bounds" business in order to better focus on density. The same reasoning led to the elimination of several variance-related items, including the limit on the number of bathrooms in structures. According to Karl C. Sollers, Seashore management assistant from 1983 to 1988 and the principal author of the 1991 zoning regulation revisions, it made little sense to count bathrooms instead of counting people, or at the very least rooms in a structure that people can occupy. "What's the difference if there is one bathroom or five? If there are twenty people in a house they will still flush the toilet the same amount of times," Sollers explained.[94] The 1991 amendments declared several variance-related items, including the number of bathrooms contained in residences and setback requirements, no longer issues of federal concern. Because the standards were made more lenient an updated review by the Secretary of Interior to determine local compliance, and whether or not the exceptions have posed a threat to the enforcement of the federal regulations, has not been deemed necessary.[95]

Nevertheless, the Seashore did not relinquish its power to condemn nonconforming structures in the Development District or anywhere else, and compliance with the zoning regulations remains a contentious issue on Fire Island. The sanction currently available to the Secretary of Interior for noncompliance is condemnation. Since actual condemnation depends upon available funding, its major strength, at present, comes from its threatened use. This "threat" has become more diluted over the years, as the average price for an interior vacant lot in many Fire Island communities has skyrocketed to $350,000 (2004 estimate). This has made the Seashore's ability to enforce compliance particularly difficult, despite outstanding zoning issues that continue to draw Seashore objections. Federal zoning standards prohibit roads and high-rise buildings and set minimum building lot sizes to prevent contiguous rows of houses. Specifically, the Federal standards set a maximum lot coverage of 25 percent for principal structures and 10 percent for accessory structures, for total maximum lot coverage of 35 percent. The standards also call for a maximum building height of twenty-eight feet. In essence, the federal standards call for improvements on no more than 35 percent of a lot, including grade-level decks and pavement. The federal zoning standards permit building on existing lots of less than 7,500 square feet, but prohibits building on new lots smaller than this. In some instances, owners of an existing six thousand square foot lot have purchased adjacent lots of the same size expecting to be able to develop it as a second parcel. The Seashore considers this just one developable parcel, since the development of a second parcel would be an expansion of a nonconforming use. Currently, in the Town of Islip, lot size is set at a minimum of 6,000 square feet and building is limited to a maximum of 30 percent of a lot for a house with a deck. In the Town of Brookhaven, a minimum buildable lot size of 4,000 square feet has been established, on which a house is limited to 35 percent coverage. In the Village of Saltaire, on the ocean side buildable parcels must be 10,200 square feet and house size is limited to 1,800 square feet, or 18 percent coverage. On the bay side, buildable parcels must be 6,800 square feet. In the Village of Ocean Beach, however, lot occupancy limits do not include improvements such as decks or patios that are under eighteen inches high. The Seashore therefore takes the position that numerous lots in Ocean Beach, the most developed area on Fire Island, are not in compliance with federal zoning standards.[96]

In the Town of Brookhaven, lots along the Great South Bay are defined as extending beyond the toe of the natural foredune line. The Seashore considers any property in front of the foredune line off limits to development. This makes for significant variations when determining the percentage of lots that are occupied or improved. The matter is complicated by the haphazard enforcement of New York State's 1981 Coastal Erosion Hazard Areas Act. The law was designed to prevent damaging construction on endangered beaches, requiring homeowners wishing to build in an "erosional hazard" zone to obtain a permit ensuring that construction will not cause any measurable change in beach erosion. However, the New York State Department of Environmental Conservation has avoided drawing a line for the coastal erosion hazard setback area on Fire Island. Fire Island is the only region of the South Shore where zoning authorities have not adopted the 1981 law, which has left the Seashore with very little support at the local level to reject building requests and prevent development.[97]

Seashore staff estimate that they reject close to 30 percent of all applications for building permits that are received for review (roughly 100 annually). The vast majority of these applications are approved by local zoning authorities anyway. Many property owners don't bother applying for permits. Federal zoning standards provide for the issuance of Certificates of Suspension of Authority for Acquisition by Condemnation when voluntarily applied for by property owners. Applications for certificates were distributed in the summer of 1992, and certificates have been distributed since 1993 for those properties which were applied for and for which eligibility was shown. The certificates do not create new protections or rights for homeowners. They merely reaffirm for homeowners the protection from condemnation afforded them if their property is in compliance with federal zoning standards. Though merely reiterating the provisions of the existing law, the idea of the certificates had been that over time only those properties without certificates would stand out as illegal, thereby encouraging compliance. Staffing shortages have prevented aggressive distribution of the certificates. Only 105 certificates have been issued as of 2004. Certificates are more popular in national seashores like Cape Cod National Seashore, where banks often require certificates before approving mortgages. The lack of appeal to homeowners on Fire Island comes from the likelihood that certificates require inspections, and Seashore staff estimate that possibly up to one-quarter of all properties in the exempt communities are nonconforming. It is also estimated that perhaps another 50 percent of properties are legally nonconforming, having been built prior to the adoption of more strict zoning codes.[98]

Mainland Terminal and Transportation

As far back as 1965, town officials and the NPS saw the need for a mainland site headquarters. At that time, Brookhaven Town had a private firm draw up plans for NPS facilities to be constructed at the Brookhaven Sandspit in Patchogue. The new headquarters building erected by the town would be rented to the Seashore and provision would be made for a transportation center for ferries servicing federal developments on Fire Island. The proposal was delayed pending the issuance of a dredging permit by the USCE. Nothing ever came of this proposal, and in 1970 the Seashore moved its headquarters to 65 Oak Street in Patchogue.[99]

Up until the time of the General Management Plan, visitors to the three existing Seashore recreation areas (Sunken Forest, Talisman, and Watch Hill), utilized private boat or commercial ferry service operating from terminals at Bay Shore, Sayville, and Patchogue. Most of the scheduled ferry service was concentrated between terminals in Bay Shore and the communities on the western portion of the Island. Ferry and special-use permits issued by the NPS to private ferry operators governed the existing ferry service from Sayville to Sunken Forest and from Patchogue to Watch Hill. Most of the other federal lands were inaccessible, except by private boat, and received little public use. Only Smith Point West is within reasonable walking distance of a parking lot (in the adjacent Smith Point County park).

Access also existed by private four-wheel drive vehicles (beach buggies), which require special permits from the Seashore and the Town of Islip or Brookhaven. These vehicles enter the Seashore from the west by way of the Robert Moses Causeway or from the east using the William Floyd Parkway. There is no continuous hard-surfaced road on Fire Island but vehicles can travel on the beach or on sand trails; however, they are not permitted to cross the Sailor's Haven/Sunken Forest area. It is the stated objective of the Seashore to reduce vehicle use on the Seashore to an absolute minimum; therefore, permits are denied whenever reasonable alternative access is available.[100]

The 1975 Draft Master Plan found that the absence of roads on the island and the existence of a major bay (Great South Bay) between Fire Island and the Long Island transportation system severely limit public access to federal lands. The unavailability of a means of ready access to the Seashore influenced visitor density and visitor use patterns. The Draft Master Plan found that visitors who did not have access to private boats may have to endure many frustrations before reaching Fire Island. Those who arrive by car often battle several hours of traffic on Long Island highways, only to become lost en route to the ferry slip, because circuitous routes were poorly marked. At the ferry slip public parking was costly and limited, leaving many visitors vulnerable to illegal parking fines. Frustration was magnified by the infrequency of ferries. If visitors missed ferry connections, waits of an hour or more were not uncommon. Some visitors came by bus or train to either the Sayville or Patchogue stations, taking taxis from there to the ferry slip. The expense and inconvenience of this method provided enough frustration to discourage most visitors. The Draft Management Plan reported that about 70 percent of the visitors to the two federal developed recreation areas (Sunken Forest and Watch Hill) arrived by private boat. On summer weekends and holidays, owners of large boats often found the marinas full. At Watch Hill, about half the slips had electrical and water hookups, and both Sunken Forest and Watch Hill had pumping stations with holding tanks. There were no dumping stations for domestic waste waters at either marina. Unlike ferryboat travelers, private boaters enter the developed areas at their leisure and avoided the temporary crowds on the docks; however, they often contended with congestion in access channels and marinas.[101]

It is therefore not surprising that the majority of visitors to the Seashore in the 1969 user survey resided in nineteen south shore Long Island communities. Others came from elsewhere in New York State, mostly from suburban counties near New York City. Only 9 percent came from the five boroughs of New York City, and 7 percent from out-of-state.[102]

With the proposed increases in public day use contained in the General Management Plan of 1978, an expansion of ferry service was envisioned. The General Management Plan recommended acquisition and development of mainland property to contain Seashore headquarters, visitor orientation facilities and ferry terminal areas. An earlier proposal contained in the Draft Master Plan proposed building a mainland facility at Blue Point at the Republican headquarters, which was discarded in 1975. A ten-acre site at the Head of the Patchogue River was thought to be well-situated in terms of proximity to public transportation, parking, arterial streets, and access to the water. In May 1976, the Seashore moved to its current location at 120 Laurel St. in Patchogue.

The Patchogue Terminal site is located very close to the Long Island Railroad station in Patchogue, thereby facilitating visitation to the park from as far away as New York City, and negating the need for an automobile, with its often-experienced traffic congestion. The headquarters is a complex of three buildings. The main building contains the administrative offices. The hindquarters building houses the park's interpretive office and library, as well as various office supplies and equipment. The third building, an old boat house in disrepair from storm damage and deferred maintenance, was moved to the Patchogue maintenance facility in December 1993 for use as a furniture maintenance building. The Patchogue maintenance facility is a complex of four buildings which houses the maintenance offices, repair shops, storage, resource management, and planning division offices. Located on the Patchogue River, the Patchogue maintenance facility also contains a marina for NPS boats.[103]

In 1977, the Seashore completed its Final Transportation Study to improve visitor access to the federal activity areas and other parts of the island. As the Draft Management Plan pointed out, Long Island is isolated from New York City and the overcrowded highway network and inefficient mass transit system impede travel easterly to the Seashore. Residents of the west end communities use four-wheel drive vehicles for access, and they are allowed to cross federal and other lands in limited numbers by special permit from both the towns (Islip, Brookhaven) and the NPS. Aside from the two federal areas developed for recreation, Sailors Haven/Sunken Forest and Watch Hill, most other federal lands were inaccessible except by private boat. Seventy percent of the visitors to these two developments arrived by private boat in 1975. This means that visitors originate primarily from Nassau and Suffolk residents.[104]

The Final Transportation Study sought to improve access to inexpensive, convenient means of public transportation to broaden the Seashore constituency. Toward that end, the Transportation Study sought modest expansions of Seashore activity areas at Sailor's Haven/Sunken Forest, Watch Hill, and Talisman. Specifically, Sailor's Haven/Sunken Forest were slated for an increase of two hundred people per day to a total capacity of twenty-five hundred daily visitors with almost two thousand arriving by ferry. Watch Hill was slated for a five hundred person per day visitation increase, with almost one thousand arriving by ferry. Talisman was scheduled for an expansion of existing day-use facilities to seven hundred persons per day, with almost five hundred arriving by ferry. To accomplish this the study called for the creation of visitor orientation facilities and two separate ferry terminals to serve Sailor's Haven/Sunken Forest, Talisman, and Watch Hill location of a mainland headquarters and ferry terminal in the Village of Patchogue and, if available, a second terminal at Heckscher State Park. More stringent control over the issuance of permits for the use of motor vehicles along the Seashore was also called for.[105]

While the proposals for a new ferry terminal at Heckscher State Park did not materialize, much of the Final Transportation Plan was implemented, including additional ferry service from Bay Shore. Today ferry service is provided by three private operators and runs seasonally from Long Island terminals at Bay Shore, Sayville, and Patchogue to

Table 5.2. Maximum Daily Visitation Levels of Existing Federally Managed Activity Areas as of June 1976

Federal Activity	*Visitors Per Day*
Sailor's Haven/Sunken Forest	2,300
Talisman	160
Watch Hill	1,500
Old Inlet	350
Smith Point West	800
Total Capacity	**5,100**

Source: Center for Regional Policy Studies

sixteen terminals on Fire Island. Route distances are from four to ten nautical miles in length. Water taxis are also available running laterally between Fire Island terminals. Currently, the network of ferry operations and Long Island and Fire Island terminals is finely tuned to the seasonal needs of island visitors (approximately 80 percent of whom come during the three peak summer months), with a primary orientation to the residents and visitors to Fire Island's communities. The operator's secondary focus are Seashore visitors. Fire Island Ferries, Inc., operating out of Bay Shore, provides service to eight communities on the west end of Fire Island. Sayville Ferry Services, Inc., operating out of Sayville, provides service on five routes to Fire Island, including service to the Fire Island communites of Cherry Grove, Fire Island Pines, and Water Island, and service to the NPS Sailor's Haven Visitor Center and Barrett Beach, both under a concession agreement with the NPS. The east end of the island, including the NPS Watch Hill Visitor Center and the community of Davis Park, is served by two Patchogue ferry terminals and the Davis Park Ferry Company and accounts for approximately 15 to 20 percent of ferry passengers. Another two cross-bay routes, one serving Point O'Woods, and the other serving Bellport Beach, are restricted to residents of Point O'Woods and Bellport, respectively. Routes and fare structures are strictly regulated by Suffolk County and a significant portion of ferry operator revenues come from parking revenues at the Long Island terminals. The Patchogue NPS ferry terminal is the only terminal with good intermodal connection to the Long Island Rail Road. Limousine and charter minibus services from Manhattan are time efficient but costly, and the LIRR/taxi link is cost effective for island community residents. Service to sites at the eastern and western extremes (Fire Island Lighthouse) of Fire Island is limited to lateral water taxi service, provided by South Bay Water Taxi and its subsidiary company, Aqualine, which may be too costly for most Seashore site visitors. There is no land-based lateral transportation service.[106]

Table 5.3. Annual Ridership by Route—2001

Mainland Departure Location	*Island Termination Location*	*Number*
Bay Shore	Kismet	89,892
Bay Shore	Saltaire	101,720
Bay Shore	Fair Harbor	89,892
Bay Shore	Dunewood	65,376
Bay Shore	Atlantique	49,032
Bay Shore	Ocean Beach	167,097
Bay Shore	Seaview	122,581
Bay Shore	Ocean Bay Park	114,089
Bay Shore Subtotal		**799,679**
Sayville	Sailor's Haven	60,500
Sayville	Cherry Grove	180,000
Sayville	Fire Island Pines	210,000
Sayville	Barrett Beach	340
Sayville	Water Island	3,000
Sayville Subtotal		**453,840**
Patchogue	Davis Park	140,000
Patchogue	Watch Hill	25,815
Patchogue Subtotal		**165,815**

Source: Fire Island National Seashore Waterborne Transportation System Plan, 2001.

Mosquito and Vector Control

The Seashore's mosquito surveillance and management program is the foundation for mosquito monitoring programs throughout the NPS. It was implemented in 1998 in response to public concern over Eastern Equine Encephalitis (EEE). In 1999, when another mosquito-borne disease—West Nile Virus (WNV)—was discovered in the New York metropolitan area, the program expanded. The goal of this program is to reduce human health risk from mosquito-borne diseases while adhering to the NPS legal mandate to protect the natural resources of the Seashore. The New York State Public Health Law of 1954 contains legal article number fifteen pertaining to insect control. The Suffolk County Bureau of Vector Control has used chemical and marsh ditching to control mosquito populations at Fire Island since the 1930s. The Seashore has formalized its working relationship with the County Bureau of Vector Control through a memorandum of understanding. Currently, the county does not implement vector control measures on federal lands and testing is conducted by Seashore personnel. This arrangement will change only if a public hazard is identified by the United States Public Health Service.[107]

The Seashore vector program focuses on the collection and analysis of salt marsh mosquitos, which are potential vectors for EEE, and various freshwater species, including the common house mosquito, which are potential vectors for WNV. Twelve mosquito traps are maintained at five different study sites in the park. The park also monitors for dead birds, vector carrier species which may have died because of WNV. Evidence of WNV-infected mosquitos has been detected within the park every year since 2000, when it was detected in the community of Saltaire. In 2001, it was detected at Watch Hill. In 2002, it was detected again at Watch Hill and in the Wilderness Area. In 2003, WNV was isolated from William Floyd Estate mosquitos, and the Seashore's first WNV positive bird was also collected in the community of Cherry Grove.[108]

Off-Road Vehicle Management

Vehicle use Seashore-wide continues to be a controversial issue for recreational and nonrecreational groups and Fire Island's residents. There have been numerous court cases against the Town of Islip and the Seashore brought by residents who claimed the Seashore did not have the right of regulation or that the regulations were harsh or discriminatory. The courts have consistently upheld the Seashore's right to regulate beach traffic.[109] President of the Fire Island National Seashore Advisory Board (FINSAB), Walter Reich, recalls that the driving restrictions put in place by Superintendent Richard Marks in the 1970s were the catalyst for the formation of his group. A set of regulations put forward by Marks in 1977, limited the residents of Fire Island to one round-trip a day by beach buggy to Long Island via the Robert Moses Causeway. During the summer, as in past years, no travel at all will be allowed. Marks often said the 1,200 to 1,300 four-wheel drive vehicles permitted on the beach during the off season up until that time caused serious coastal erosion.[110] The FINSAB, made up of about five thousand local residents, joined with the United Mobile Sportsfishermen sports clubs, fishermen, and businesses to lobby Washington, D.C. and the Seashore on a number of issues important to residents, but none more pressing to the group than the the need to have access to the Seashore via automobile.[111] Ultimately, through discussions, the driving rules were amended in a way that categorized driving permit holders by function and created a minimum and maximum number of permits for each category. Federal regulations governing motor vehicle use on Fire Island

(36 CFR 7) date back to 1968 and were amended in 1977 (42 FR 62483), 1984, and 1987. More recently, the Seashore's system of allocating driving permits has come under criticism. The current vehicle-use regulations were established in 1987 as part of the original regulations authorizing the Seashore Superintendent to "limit the total number of permits for motor vehicle travel on Seashore lands, and/or to limit the number of permits issued for each category of eligible applicants." The existing permit program is outlined in Section (a) (8) of 36 CFR 7.20. This section specifies the limits on vehicle permits and travel for Fire Island residents, governments, businesses and recreational users. Specifically, the regulations establish the following limitations.

- Year-round resident permits will be limited to 145; they are good for one year and only one permit per household will be issued.
- Two round trips per day for year-round residents are permitted.
- Part-time resident permits are limited to those who either held year-round or part-time resident permits as of January 1, 1978.
- Permits will be issued for those with retained rights of use and residency proven by deed.
- Public utility and essential service vehicles will be limited to 30 permits per year.
- Construction and business vehicles will be limited to 80 permits at any one time.
- Up to 5 permits for each community for the use of year-round municipal employees will be permitted.
- There is a 5000 trip/year limit on recreational vehicle trips.
- The seashore is closed to recreational vehicle travel from January 1 to March 31 and June 14 to September 14.

The regulations also specify seven permit user groups: Year-Round Residents; Part-Time residents; Reserved Rights of Use and Occupancy; Construction Permits and Business Vehicles; Municipal Employees; and, Recreation Permits. Special-use permits have been granted to provide beach access to businesses and public utilities performing personal services for residents of Fire Island and to commercial operators of government-owned facilities. Ferry operators and water taxi services have permits for docking and loading and unloading passengers at government facilities. Public utility permits are multiyear. Others are negotiated with expiration of each concession contract.[112]

At various times year-round residents of Fire Island have protested against the granting by the Seashore of driving permits to part-time residents, who the Seashore believes should not be eligible to drive on Fire Island. The towns of Islip and Brookhaven, which split jurisdiction over Fire Island, make the initial permit recommendations, and the Seashore makes the final decision. After the first year a person is issued a permit, it is issued annually by mail. But there is a waiting list for driving permits for year-round residents, and distinguishing between part-time and year-round residents is often difficult. The permit for driving applications asks: "Are these quarters your sole year-round domicile?" A booklet of guidelines says year-rounders must live in their Island homes "continuously, except for brief and occasional absences, for 12 months of the year." The chief ranger of the Seashore has the authority to issue the federal permits. The objective of the Seashore in distributing permits has not been to make it convenient for homeowners to come out to Fire Island on weekends. However, the Seashore does not commit to a clear definition of how many nights a week year-round residents are permitted to live off-island. For the most part, year-round residency is determined by the Islip Driving Commission through a review of the applicant's driver's license, vehicle registration, insurance statements, electricity bills and itemized phone bills. According to administrators of the current system, "only a limited number of applications are denied because the applicants are found not to be year-round residents."[113]

In 1998, the Superintendent of the Seashore initiated an internal review of the regulations governing motor vehicle use on Fire Island National Seashore lands. In response to concerns of local residents, visitors, and Seashore staff, the Superintendent sought to explore the possibility of building consensus on new regulations controlling motor vehicle use within the Seashore. To assist the Seashore in an assessment of driving on Fire Island and to ensure objectivity of the assessment, the Superintendent retained the Consensus Building Institute (CBI) to conduct this Conflict Assessment. CBI is a nonprofit organization specializing in consensus-based solutions. The result of this effort was the Negotiated Rulemaking Advisory Committee for Off-Road Driving Regulations at Fire Island National Seashore, which resulted in the following regulations: the Seashore will issue three public utility permits (one each for water, phone, and electricity); up to 145 year-round resident permits; no more than 100 part-time resident permits (with the intention of reducing this number to zero over time); up to thirty "essential service" and municipal or community employees permits; up to one hundred contracting and business permits; and up to ten special permits.[114]

Personal Watercraft

Since 2002, as the result of a lawsuit, personal watercraft have been prohibited in all NPS waters, including Fire Island National Seashore. This was the result of a legal challenge over the effects of personal watercraft on other visitors, their safety, water and air quality, submerged and shoreline vegetation, and wildlife and its habitat. In 2004, new regulations are scheduled to go into effect that will allow personal watercraft in some Seashore waters. But personal watercraft will be prohibited within one thousand feet of shore in the Great South Bay from Robert Moses State Park to the western boundary of Sunken Forest and within 4,000 feet of shore in the Great South Bay from the western boundary of the Sunken Forest to Smith Point County Park.[115]

Management and Organizational Structure of the Seashore

The Fire Island National Seashore continues in its mission to preserve the natural and cultural resources within its administrative boundaries. To do this the Seashore maintains an annual operating budget of $3.4 million, which supports a work force of fifty permanent positions, two term positions, and fifty seasonal positions. Park staff is organized into five operating divisions: Resource Management, Interpretation, Visitor Protection, Maintenance, and Administration. The current personnel and organizational structure of the Seashore has not changed dramatically since its inception in 1964. See figures 5.1 and 5.2.

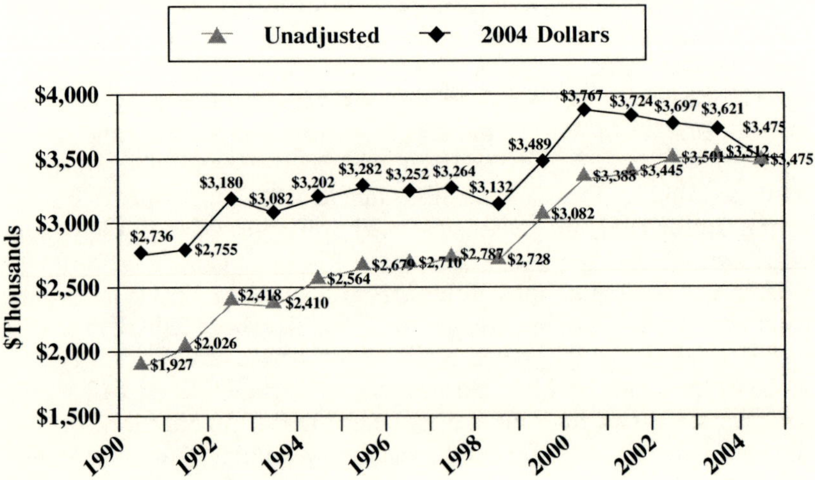

Source: Center for Regional Policy Studies

Figure 5.1. Total Budget Fire Island National Seashore Unadjusted and Adjusted

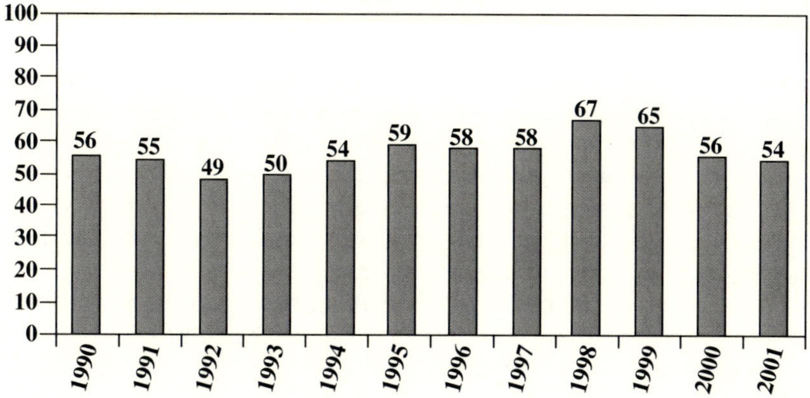

Source: Center for Regional Policy Studies

Figure 5.2. Total Staffing Fire Island National Seashore Full Time Equivalent Employees

The perceived staffing needs of the Seashore, as envisioned by the North Eastern Regional Director of the National Park Service in 1963, provided for 51.5 full-time equivalent employees (FTEs) and a budget of $2.2 million. This number was to grow every year for the first five years until FTEs reached 123.5 by the fifth year of Seashore operations. In reality, staffing levels did not exceed fifty FTEs until the 1980s, and by 1974 there were only 48.2 FTEs. The original plan for the Fire Island National Seashore called for a management unit with three employees, and an administrative unit with five employees, a ranger unit with seventy employees, a nature unit with five employees, and a maintenance unit with seventy-six employees.

Under the NPS's area cluster management Fire Island National Seashore and the New York City National Park Service Group were merged into a single administrative unit late in 1966. A GS-15 superintendent position was established to oversee the grouping and was filled by Hank Schmidt. Simultaneously a new position of assistant superintendent was established to administer Fire Island, filled by Thomas F. Norris, Jr. In June 1967, a Management Appraisal Team visited Fire Island and among the things they recommended was the reestablishment of the Seashore as an independent management unit. In January 1971, James W. Godbolt became assistant superintendent of Fire Island National Seashore. In the spring of that year, Mr. Godbolt's title was subsequently changed to Superintendent. He recommended certain organizational changes in the staff at Fire Island. In February 1972 the area was divided into two units: East and West. Each was headed by a Unit Manager. All maintenance and protection and maintenance and rehabilitation functions were placed under the respective unit managers giving each the opportunity to plan and program a complete park operation. Under the reorganization, two maintenance supervisors and two supervisory ranger positions under the unit managers were created. After another management appraisal team visit in 1973, they recommended that the unit manager form of organizational operation be changed back to staff and line, believing there were too many layers of supervision. This was done in 1973. In 1976, the organizational makeup of the Seashore was changed again to a direct line operation.[116]

By 1974, the Seashore consisted of a superintendent's office, under which a chief of operations served. The chief of operations was responsible for overseeing a unit manager for the western part of the Seashore, a repair shop, an interpretation unit, and a unit manager for the eastern part of the Seashore.

The General Management Plan of 1978 divided the Seashore into six management units: the Fire Island Lighthouse management unit; Sunken Forest management unit; Watch Hill management unit; the high dune management unit, Moriches area (Suffolk County lands managed by Suffolk Count Parks Department), and the William Floyd management unit. After the creation of the Otis Pike High Wilderness Area in 1980, this organization was amended to reflect seven management units: the Fire Island Lighthouse Management Unit; the Sunken Forest Management Unit; the Wilderness Management Unit; the William Floyd Management Unit; the Patchogue Village Unit; the Development Unit; and the Exempted Communities.

From 1982 through 1987, the organization of the Seashore reflected the unit structure laid out in the General Management Plan. The park superintendent oversaw an administrative unit with five employees (the Patchogue Village Unit), a William Floyd Estate unit with three employees (the William Floyd Unit), a Division of Interpretation and Resource Management which included supervising park rangers, an East District covering the Wilderness Management Unit, a West District covering the exempted communities unit, and an Interpretation Unit covering the Development Unit, the Sunken Forest Unit, and the Fire Island Lighthouse Unit. The Seashore also included a large maintenance unit, with twenty-three employees.

This has been the core of the Seashore's organizational structure up until the present time, though there has been some shuffling of units to reflect the Seashore's growing concern with natural resource preservation and protection. In 1988, the Seashore's Organization Review proposed divisions named the Division of Visitor Services and Resource

Protection, a Division of Planning and Resource Management, a Division of Administration and a Division of Maintenance.[117] The organizational chart for 2003 indicates that some of these recommendations have been implemented. The 2003 charts show a superintendent's office with the following positions: Superintendent, Deputy Superintendent, Landscape Architect, Superintendent's Secretary. The Superintendent and the Deputy Superintendent oversee the William Floyd Estate Unit, the Division of Interpretation, the Division of Administration, the Division of Visitor and Resource Protection, the Division of Maintenance, and the Division of Resource Management.

In 2004, along with the superintendent, staff expertise, and specialities include fifteen permanent park rangers, eleven permanent maintenance workers, three biologists, a curator, an environmental protection specialist, and a park planner. There are seventeen seasonal/temporary park rangers, with twelve working in interpretation, and six in visitor protection. There are nine seasonal maintenance workers. The superintendent's and administrative staff total eight. This work force is supported by approximately twelve thousand hours of Volunteers-in-Parks service, six student conservation assistants, and special project and program funds distributed by the NPS's regional and Washington offices.[118]

Seashore funding also goes to support three visitor centers, one campground, twenty-one miles of beaches, seventeen miles of hiking trails, one central and three auxiliary maintenance facilities, main and auxiliary park headquarters buildings, twenty-three employee housing units for on-site protection and management of park resources, and one ferry terminal and parking lot.[119]

The Seashore has recently gone through some leadership changes. Constantine Dillon, the Superintendent from 1997 to 2003, received high marks from environmentalists but clashed often with Fire Island residents over many issues. Fiercely committed to maintaining the Seashore in as natural a state as possible, Dillon was aggressive in blocking homeowner backed beach replenishment projects, enforcing the ban on personal water craft, and enforcing beach closures to protect the Piping Plover during hatching season. Mr. Dillon, who believed that park resources were not threatened by wave action, also opposed the USCE's Interim Project to replenish dunes in undeveloped areas of the Seashore.[120] Dillon left the Seashore to head up the NPS training center at Grand Canyon National Park in 2003. The move was widely seen as a promotion for Dillon, but some Fire Island residents believed their agitation and lobbying played a role in his leaving.[121]

Dillon was replaced in May 2003 with David Spirtes, a Bronx native who has been credited with improving relations with Fire Island residents. Spirtes died after a bout with cancer after less than a year at

Table 5.4. Organizational Chart and Budget Comparison

Division	*Staff*	*5th Year*	*1963*	*1974*	*1982*	*1987*
Management	PFT	3	3	2	3	2
	PPT		1			1
Subtotal		3	4	2	3	3
Administration	PFT	5	3	3	4	5
	SFT			1	1	
	PPT			1	1	2
Subtotal		5	3	5	6	7
Ranger	PFT	25	4	11	11	17
	SFT		4	2	2	6
	PPT		1			
	TEMP	45	55	30	20	27
Subtotal		70	64	43	33	50
Planning and Resource Mgt	PFT					6
	STF					1
Subtotal						7
Naturalist	PFT	3	1	3	3	8
	STF				2	3
	TEMP	2	5	9	13	15
Subtotal		5	6	12	18	26
Maintenance	PFT	70	7	11	11	17
	STF		5	5	12	10
	PPT		1			
	TEMP	6	15	17	22	15
Subtotal		76	28	33	45	42
Grand Total		159*	105	95	105	135
FTE		123.5**	48.2	60.3	51.3	93.15
Total Budget (Absolute $–000)		$1,128	$ 765	$1,464	$1,799	$3,100
Total Budget (1988 $–000)		$4,377	$1,752	$2,577	$1,800	$3,100

Note: *Does not include 5 positions—Acquisition Project Mgr Office/**Estimated-average employment periods

the helm, but by all accounts he succeeded in that short period in changing the tone on Fire Island. "He changed the atmosphere overnight," said Gerald Stoddard, president of the Fire Island Association. "It had been one of contention and mistrust, and he was the type of person who demonstrated what could be done with cooperation." Barry Sullivan, a deputy to the superintendent, served as acting superintendent immediately after Spirtes's death, followed by Dave Reynolds in 2004. A new superintendent will be chosen by the NPS, with substantial input from Marie Rust, the regional director of the NPS's northeast region.[122] To date, the Seashore has acquired over 3,000 acres in fee and 3,100 in easements. With the exception of 12.51 acres of land acquired within communities, fee lands acquired represent property needed for public use and environmental protection. The easement lands represent beach face and underwater properties.[123] Table 5.5 is a breakdown of land ownership on Fire Island today.

Appendix A is a list of the Fire Island Superintendents. Appendix B lists a chronological timeline from 1964–2004 of Fire Island National Seashore highlights.

Figure 5.5. Land Ownership—2004

Federal		
	Fee	3,082.99 acres
	Less than fee	3,151.51 acres
	Total Federal	**6,234.50 acres**
Non-Federal		
	Private	918.92 acres
	Public	12,425.13 acres
	Total Non-Federal	**13,344.05 acres**
Park Total		**19,578.55 acres**

Source: Center for Regional Policy Studies

Appendix A

Fire Island Superintendents

Henry Schmidt	1964–1966
Thomas Norris	1966–1971
James Godbolt	1971–1975
Richard Marks	1975–1980
Joel Pickelner	1980–1982
John Hauptman	1982–1987
Noel Pachta	1987–1991
John Hauptman	1991–1996
Constantine Dillon	1997–2003
David Spirtes	May 2003–April 15, 2004
Barry Sullivan (acting)	May 2004–November 2004
Dave Reynolds (acting)	November 2004
Michael T. Reynolds	December 2004–present

Appendix B

Fire Island Timeline 1964 to present

1963: *Proposed Fire Island National Seashore, Suffolk County, New York*. CRBIB Number: 401146. Holdings: NESO.

Supplementary Report on Proposed Fire Island National Seashore, Suffolk County, New York. CRBIB Number 401147. Holdings: NESO.

1964: President Lyndon B. Johnson signs into law the creation of Fire Island National Seashore "for the purpose of conserving and preserving for the use of future generations certain relatively unspoiled and undeveloped beaches, dunes, and other natural features . . . which possess high values to the nation as examples of unspoiled areas of great natural beauty in close proximity to large concentrations of urban population."

Fire Island joins four other national Seashores already in existence: Cape Hatteras, North Carolina; Cape Cod, Massachusetts; Padre Island, Texas; Point Reyes, California.

Former Secretary of the Interior, Stewart L. Udall (1960–1968) officiates at the official dedication ceremony on Fire Island in 1964.

October opening of an office at the Sayville Motor Lodge in Sayville, New York.

$16 million allocated for land acquisition.

Meeting held with President of the Sunken Forest Preserve, Inc., James N. Dunlop, to discuss donation to the National Park Service of the seventy-five acres of the Sunken Forest property that the group holds east of Point O'Woods.

Federal Wilderness Act of 1964.

June:	Bridge built over the Fire Island Inlet to provide access to newly christened Robert Moses State Park
	NPS properties located in the New York metropolitan area grouped into one administrative unit known as the New York City Group (NYC Group), Headed by John A. Townsley. Included originally: the Statue of Liberty, Federal Hall, Castle Clinton, the General Grant and Hamilton Grange National Memorials, and the Theodore Roosevelt Birthplace and Sagamore Hill National Historic Sites.
	Henry G. Schmidt named management assistant for Seashore.
1965:	Draft Master Plan for Fire Island National Seashore completed.
March 10:	First direct payment for Seashore lands accomplished with the purchase of the .183 acre Albert Brower Tract near Whalehouse Point.
May:	Groundwork for oversand vehicles regulation laid at conferences held with representatives of the Great South Beach and Long Island Buggy Associations.
	Legislative amendment authorizes the Secretary to accept donations of lands and structures at the William Floyd Estate for inclusion as a detached unit of the Seashore.
July:	Suffolk County Board of Supervisors approves the construction of the Smith Point Visitor Contact Station by the NPS. Construction of three prefabricated geodesic dome coverings.
	Talisman acquired.
	First eleven groins constructed by USCE east of Moriches Inlet as part of 1959 Reformulation Plan.
	Additions in all categories of the parks system aided by the Land and Water Conservation Fund Act of 1965.
April 2, 1966:	Zoning standards established. Required that towns and villages submit their zoning ordinance to the Secretary for approval.

April 14: Sailor's Haven acquired.

May: The Sunken Forest acquired.

May 27: Secretary of Interior Stewart Udall breaks ground at Watch Hill to mark the start of construction of the first public facilities within the boundaries of the Seashore.

National Historic Preservation Act signed into law.

1967: Seashore added to the NYC Group, changing the Group's name to the Fire Island National Seashore and New York City NPS Group.

Henry G. Schmidt leaves to run NYCNPS Group.

Thomas Norris named Superintendent of the Seashore.

July: Watch Hill 166 boat marina completed.

Joint Seashore, Town and Village Committee chaired by Suffolk County Commissioner of Parks Charles R. Dominy lay groundwork for uniform revisions of local oversand vehicle regulations.

1968: Some homeowners negotiate twenty-five year period for occupation of structures in unauthorized section of Seashore.

Old Inlet and Fireplace Beach Club acquired.

Fire Island hit by severe hurricane.

Federal regulations governing motor vehicle use on Seashore are adopted.

1969: Visitor survey finds majority of visitors from Nassau-Suffolk region. Only 9 percent from NYC.

August: The Town of Islip receives approval of an amended zoning ordinance from the Secretary.

Watch Hill marina handed over to concessioner.

Two groins added to original eleven east of Moriches Inlet by USCE from 1959 Reformulation Plan.

Fire Island hit by severe hurricane.

1970: National Environmental Policy Act of 1970 requires completion of management plan. Work on master plan begins.

Seashore moves headquarters from Sayville to 65 Oak Street in Patchogue.

Watch Hill concession building completed.

1971: James Godbolt named superintendent of Seashore.

Oversand vehicles regulations amended.

Draft Master Plan work continues.

1972: Fourteen property owners file suit against the federal government on the grounds that the Department of the Interior was not controlling development. Property owners defeated.

Pump out station installed at Watch Hill marina.

Endangered Species Act passed by Congress.

Draft Master Plan work continues.

February 1973: *Conceptual Master Plan For Fire Island National Seashore, Suffolk County, New York*, completed by Denver Service Center. CRBIB Number: 401156. Holdings: NESO.

Facility with an office, exhibit area, first-aid and locker room, marine storage, restrooms and showers, grocery store and marine shop, and lifeguard locker room completed at Watch Hill.

The Flood Disaster Protection Act of 1973 (P.L. 93–234) requires that every community identified by U.S. Department of Housing and Urban Development as having areas of special flood hazard participate in the National Flood Insurance Program. Fire Island has been identified as having special flood hazards and approximately 80 percent of the Island's residences are covered by this federally subsidized insurance.

The New York State Tidal Wetlands Act of 1973 passes. Regulates the use and development of salt marshes and adjoining areas. South Shore Estuary Reserve Council established by the Governor to monitor the estuary and its watershed, establish measures to protect, preserve and restore its resources, and evaluate its significance for inclusion into the National Estuary Program.

Historic Resource Study, The William Floyd Estate, Mastic, Long Island, by Ricardo Torres-Reyes completed for U.S. Department of the Interior, National Park Service, New York District. CRBIB Number: 011919. Holdings: WASO/NESO/FIIS.

1974: 150 years of lightkeeping on Fire Island ends when lighthouse is replaced by a strobe light on the nearby Robert Moses State Park water tower.

October: *Report of an Archeological Reconnaissance of Fire Island, Suffolk County, New York*, by John F. Vetter and Bert Salwen. Chadwyck-Healey. CRBIB Number: 012726. Holdings: HFC/FIIS/NESO.

December: Babcock and Roddewig produce *Report to the National Park Service on Fire Island National Seashore.*

Archeological reconnaissance finds no premodern remains on Fire Island.

Historic Resource Study, The William Floyd Estate, Fire Island National Seashore, New York, by Ricardo Torres-Reyes completed for the U.S. Department of Interior, National Park Service. CRBIB Number: 000435. Holdings: WASO/HFC/CRC/FIIS.

1975: Richard Marks named superintendent of Seashore.

Draft Environmental Impact Statement, Master Plan for Fire Island published.

Draft Master plan proposal to establish headquarters at Republican headquarters in Blue Point discarded.

Management plan concept replaces master plan concept.

February: Senator Jacob Javits introduces bill to authorize additional $10 million for acquisitions of property within the dune district zone. Only $2 million authorized.

Environmental Inventory of the Fire Island National Seashore and the William Floyd Estate, Suffolk County, New York, by Jack McCormick & Assoc., Inc., completed. CRBIB Number: 401166. Holdings: NESO.

November: *Fire Island National Seashore: Preliminary Report of Findings and Recommendations*, by William K. Reilly. CRBIB Number: 401154. Holdings: NESO.

June 1976: *Draft General Management Plan Summary, Fire Island National Seashore, New York* completed. U.S. Department of Interior, National Park Service. Fire Island General Management Plan completed. CRBIB Number: 401153. Holdings: NESO

May: Seashore moves to its current location at 120 Laurel St. in Patchogue.

August: Meeting of 400 Fire Island residents against the *Draft General Management Plan's* proposal for a Dune District.

Environmental groups, residents and governmental officials support *General Management Plan.*

Congress raises acquisition cost ceiling to $18 million.

Mainland Feasibility Study: Patchogue Terminal, Fire Island National Seashore, by Vollmer Associates, U.S. Department of the Interior, National Park Service. CRBIB Number: 401140. Holdings: NESO.

Public use of William Floyd Estate still negligible.

October: Floyd descendants relinquished their use and occupancy rights to 33.8 acres of the estate, including the major residence and outbuildings.

Dune District designated by a dune crestline and mapped. Extends from the mean high water line on the beach northward to forty feet behind that mapped crest of the primary dune.

1977: *General Management Plan, Fire Island National Seashore,* completed. U.S. Department of the Interior, National Park Service. CRBIB Number: 401152. Holdings: NESO/CRC.

September: *Final Environmental Statement/General Management Plan, Fire Island National Seashore, New York.* U.S. Department of the Interior, National Park Service. CRBIB Number: 401151. Holdings: NESO.

Deadline for squatters of Labor Day passes without squatters vacating. February 15, 1978 set as new date.

Exploring the Sunken Forest promotional brochure published.

Administrative History of Fire Island National Seashore by Betty Thompson published.

Seashore reaches acquisition of 5,944 acres of land in fee or easement within the Seashore's boundaries.

Federal Regulations governing motor vehicle use on Fire Island National Seashore amended.

Final Transportation Study, Vollmer.

May: *Final Report on a Document Review and Archeological Reconnaissance of the William Floyd Estate and the Fire Island Lighthouse Tract, Fire Island National Seashore, Suffolk County, New York,* completed. CRBIB Number: 012727. Holdings: NARO/HFC/CRC.

Drawing to Accompany Historic Structure Report, The William Floyd House, Fire Island National Seashore, Mastic, Long Island, New York, by John M. Dickey completed for the U.S. Department of the Interior, National Park Service. CRBIB Number: 010350. Holdings: WASO/CRC. Available from Chadwyck-Healey.

Historic Structure Report, The William Floyd House, Fire Island National Seashore, Mastic, Long Island, New York, by John M. Dickey. CRBIB Number: 010348. Holdings: WASO/NESO/CRC/FIIS. Available from Chadwyck-Healey

Index for Historic Structure Report, Volume I, The William Floyd House, Fire Island National Seashore, Mastic, Long Island, New York, by John M. Dickey completed for the U.S. Department of the Interior, National Park Service. CRBIB Number: 010349. Holdings: WASO/CRC/FIIS.

1978: General Management Plan is adopted.

Original act creating the Seashore is amended to include within its boundaries bay islands. Suspends powers of Secretary to condemn property in Dune District as long as owner(s) keep property in natural state. Includes within Seashore boundary New York State and U.S. Coast Guard properties in the vicinity of the Fire Island Lighthouse and twelve acres of land on the Patchogue River in Patchogue for the new Seashore headquarters/terminal.

Spending authority for acquisitions is increased to $23 million, though amount not appropriated.

Assessment of Alternatives: William Floyd Estate, Fire Island National Seashore, U.S. Department of the Interior, National Park Service. CRBIB Number: 401149. Holdings: NESO/CRC.

Off-road vehicle regulations completed.

August 2: Twenty-one of the squatter families take action as plaintiffs against Seashore Superintendent Marks and Secretary of the Interior Cecil B. Andrus in the United States District Court for the Eastern District of New York.

September 13: Judge John R. Bartels of the Federal District Court ruled that the NPS was within its rights to evict these squatters.

Squatters evicted after summer season.

Construction prohibited in Dune District, except for necessary access crossings.

The USCE files an Environmental Impact Statement with the United States Environmental Protection Agency (EPA) relative to completion of the original reformulation project. The filing is rejected by Council of Environmental Quality, mostly because of concern over the proposed use of heavy structural devices such as groins to sustain inlets.

Dune District mapped in 1976 is adopted by Congress.

Interpretive Prospectus: Fire Island National Seashore. CRBIB Number: 401143. Holdings: NESO.

December 8, 1979: Richard Marks replaced by Acting Superintendent William Schenk.

Historic Resources Study completed. Identified fifteen zones in the Seashore with historically significant resources, including life saving stations, the fish factory, Camp Cheerful, historic homes, and more.

Draft Statement for Management completed.

Investigation to determine the feasibility of wilderness designation completed in fall.

An investigation to determine the feasibility of wilderness designation completed.

Fire Island National Seashore Historic Resource Study: Interviews, by Laraine A. Fletcher and Ellen R. Kintz. CRBIB Number: 401150. Holdings: NESO.

Fire Island Project: Documentary Research, Preliminary Report No. 1, by Laraine A. Fletcher and Ellen R. Kintz. CRBIB Number: 400137. Holdings: NARO/FIIS.

Historic Resource Study, Fire Island National Seashore, Long Island, NY 2 vols, by Laraine A. Fletcher and Ellen R. Kintz, U.S. Department of the Interior, National Park Service. CRBIB Number: 010346. Holdings: WASO/NESO. Available from Chadwyck-Healey.

Preliminary Report No. 2: Cultural Resources Study, Fire Island National Seashore, Patchogue, New York, by Laraine A. Fletcher and Ellen R. Kintz. CRBIB Number: 400078. Holdings: NARO/FIIS.

Preliminary Report No. 3, Cultural Resource Study: Fire Island National Seashore, Patchogue, New York, by Laraine A. Fletcher and Ellen R. Kintz. CRBIB Number: 400076. Holdings: NARO/FIIS.

Legislative History of Fire Island National Seashore, U.S. Department of the Interior, National Park Service. CRBIB Number: 401174. Holdings: NESO/FIIS.

Description of Selected Alternative Site Development and Facilities: William Floyd Estate, Fire Island National Seashore, U.S. Department of the Interior, National Park Service. CRBIB Number: 401148. Holdings: NESO.

January 1980: *Fire Island National Seashore, N.Y.: Wilderness Study/ Preliminary Wilderness Proposal and Draft Supplement* *to the Final Environmental Statement General Management Plan.* U.S. Department of the Interior, National Park Service. CRBIB Number: 401155. Holdings: NESO.

Congress creates the Otis Pike Fire Island High Dune Wilderness legislation encompassing 6.5 miles from Watch Hill on the west to Smith Point County Park on the east.

Zoning standards for Seashore amended.

October 15: William Floyd Estate entered in the National Register of Historic Places.

William Floyd Estate named separate unit of Seashore with unit manager reporting directly to Superintendent.

Joel Pickelner named superintendent of Seashore.

Reformulation of erosion control project initiated by USCE.

Stabilization of Main Building, William Floyd Estate, Fire Island National Seashore, New York, by Meadows/Woll Architects. CRBIB Number: 010347. Holdings: WASO/CRC/FIIS.

1981: New York State's 1981 Waterfront Revitalization and Coastal Resources Act, which created Coastal Erosion Hazard Areas. Not adopted by zoning authorities on South Shore.

Judge Eugene H. Nickerson, of Federal District Court in Brooklyn, upholds Seashore's power to condemn properties owned by Fredric Rose in Fire Island Pines, a decisive victory for federal zoning standards.

Archeological Investigation for Structural Stabilization of the William Floyd Estate Manor House, Fire Island National Seashore. CRBIB Number: 013296. Holdings: WASO.

Corporate charter of the Fire Island Voters Association established in 1961 is modified and renamed the Fire Island Association. The FIA Board of Directors comprises the mayors of the villages of Ocean Beach and Saltaire and the leaders of eighteen community associations. The Board elects a president, first vice-president, treasurer, secretary and three vice-presidents every two years.

Final Report on Effects of Off-Road Vehicles on Beach & Dune Dynamics, Leatherman.

William Floyd Estate opens to the public, first floor only of main house.

1982: Fire Island Lighthouse Preservation Society formed.

The Effects of Structures on Dune Dynamics, report submitted.

John Hauptman replaces Joel Pickelner as superintendent of Seashore.

Second floor of main house, William Floyd Estate opens to the public.

1983: Madeleine C. Johnson publishes *Fire Island: 1650s–1980s* (Mountainside, New Jersey: Shoreland Press, 1983).

Historic Structure Report, Architectural Data Section, the Fire Island Lighthouse and Keeper's Dwelling, Fire Island National Seashore, Patchogue, New York, U.S. Department of the Interior, National Park Service, North Atlantic Region, North Atlantic Historic Preservation Center. CRBIB Number: 012913. Holdings: WASO/NESO/CRC/FIIS.

William Floyd Estate: Room Guides for Interpretation, by Steven Kesselman. CRBIB Number: 401171. Holdings: NESO.

April: *Tri-Racial Isolates in a Bi-Racial Society: Poospatuck Ambiguity and Conflict*, by Ellice B. Gonzales. CRBIB Number: 401163. Holdings HFC. Available from Chadwyck-Healey.

September: *Final Report, 1981 Archeological Investigations, Fire Island National Seashore, William Floyd Estate, Long Island, New York*, by Dana C. Linck, completed for the U.S. Department of Interior, National Park Service. CRBIB Number: 011917. Holdings: HFC/NESO/CRC/FIIS.

Wilderness Management Plan, Fire Island National Seashore. CRBIB Number: 401161. Holdings: NESO.

Field Management Plan, Reclamation of Fields, William Floyd Estate, by Richard Stavdal and Allan O'Connell. CRBIB Number: 401168. Holdings: NESO.

February 1984: *From Unkechaug to Poospatuck*, by Ellice B. Gonzales. CRBIB Number: 011918. Holdings HFC Available from Chadwyck-Healey.

Land Protection Plan completed.

Powderpost Beatle monitoring program begins at William Floyd Estate.

Fire Island Lighthouse placed on the National Register of Historic Places.

Fire Island National Seashore Amendments Act of 1984 amends sections 459e–1 and 459e–2 to acquire private properties within Seashore boundaries. Authorizes superintendent to sell land acquired to lowest bidder or retain monies from sale in order to purchase more land.

Federal regulations governing motor vehicle use on Seashore amended.

Collections Conditions Survey, NARO

1985: Secretary of Interior approved compliance with federal zoning standards for the four zoning authorities within the Seashore.

Fire Island Inlet to Montauk Point Long Island, New York: Reformulation Study published by National Park Service for USCE.

Barry Mackintosh publishes *The National Parks System: Shaping the System,* 5th ed. (Washington, DC: National Parks Service, 1985).

Fire Island hit by severe hurricane.

May: *Final Report, Assessment Phase of the Archeological Collections Accountability Project*, by Linda A. Towle completed for U.S. Department of the Interior, National Park Service, North Atlantic Regional Office, Division of Cultural Resources, Eastern Archeological Field Laboratory. CRBIB Number: 403864. Holdings: NESO.

1986: In preparation for its relighting Seashore constructs universal access walk at Fire Island Lighthouse Beach with donated materials and labor provided by Friends of Lighthouse Beach.

February: *Vegetation and Land-Use History of the William Floyd Estate, Fire Island National Seashore, Long Island, New York*, by James S. Clark, U.S. Department of Interior, National Park Service, North Atlantic Region. CRBIB Number: 011918. Holdings NESO/OCLP.

May 28: Memorial Day. Fire Island Lighthouse relit and reinstated as an official aid to navigation.

Piping Plover (*Charadrius melodus*), Least tern (*Sterna antillarum*), Roseate tern (*Sterna dougalli*), Seabech Knotweed (*Polygonum glaucum*), and Seabeach Amaranth (*Amaranth pumilus*) come under federal monitoring under terms of the Endangered Species Act.

March: *Report of Archeological Testing at the William Floyd Estate, Fire Island National Seashore*. Rockville, MD: U.S. Department of the Interior, National Park Service, Denver Service Center, Eastern Applied Archeology Center. CRBIB Number: 400077. Holdings: NARO/HFC/FIIS.

Archeological Testing at the Fire Island Lighthouse Tract, Fire Island National Seashore, New York, by Paula A. Zitzler completed for U.S. Department of the Interior, National Park Service, Denver Service Center, Eastern Applied Archeology Center. CRBIB Number: 013803. Holdings HFC/NESO/FIIS.

August: *Archeological Testing at the Fire Island Lighthouse Tract, Fire Island National Seashore, New York*, by Paula A. Zitzler completed for U.S. Department of the Interior, National Park Service, Denver Service Center, Eastern Applied Archeology Center. CRBIB Number: 013803. Holdings HFC/NESO/FIIS.

1987: Construction was completed on a parking lot and associated road system at William Floyd Estate.

Museum of Natural History duplicates the original J. T. Nichols journals at no cost to the NPS.

External recoating and painting completed at the Fire Island Lighthouse, completing Phase III of a four-phase project.

Noel Pachta replaces John Hauptman as superintendent of Seashore.

Passage of amendments to driving permit regulations.

Evaluation of Humidity/Heating Problem: William Floyd Estate, Fire Island National Seashore, Title I Study Report, by Michael Lynn & Associates. CRBIB Number: 401172, Holdings: NESO/FIIS.

1988: Over 190 documents dating back to the 1680s donated to the estate by a distant relative of the Nicholls family.

$5,500 was made available from the Fire Island Lighthouse Preservation Society to help keep the Keepers Quarters Museum open during the summer.

Patchogue Ferry Terminal Site Development Plan, NARO.

Archeological Investigation, 1978–1986, Fire Island National Seashore, William Floyd Estate, Long Island, New York, by Dana C. Linck and L.D. Meyers completed for U.S. Department of the Interior, National Park Service, Denver Service Center. CRBIB Number: 403376. Holdings CRC/FIIS.

1989: Fire break (clearing of vegetation) constructed around part of the William Floyd Estate to address a high incidence of arson.

Fifteen of eighteen historic fields at the William Floyd Estate reclaimed.

February 15: WLIM signs historic lease for use of the Lighthouse tower. Check for $100,000 was presented to the NPS.

Crushed concrete surface was added to the Lighthouse road.

Watch Hill Visitors Center was rehabilitated through Harpers Ferry Center.

Fire Island deer herd approaches 500.

Long Island Regional Planning Board publishes *Proposed Long Island South Shore Hazard Management Program* (Hauppauge, New York: n.p. 1989).

1990: Ferry Terminal Site Project Manual completed, Denver Service Center

Fire Island National Seashore, Collection Storage Building, Design Proposal Alternatives, James A. Cohen and Alan Cerny, Albany, NY: Mesick Cohen Waite Architects. CRBIB Number: 403369. Holdings CRC/FIIS.

Cultural Landscape Report, NARO. William Floyd Estate Interpretive Training Guide. CRBIB Number: 401170. Holdings: NESO/FIIS.

Report of the Collection Accountability Project at Fire Island National Seashore, Charlestown, MA, by Will Jeffers and Jule Ratner. CRBIB Number: 403892. Holdings: FIIS.

1991: Seashore zoning standards amended.

The 1991 amendments declare several variance-related items, including the number of bathrooms contained in residences and setback requirements, no longer issues of federal concern.

Twenty-five year lease for family on William Floyd Estate expires.

John Hauptman replaces Noel Pachta as superintendent of Seashore. It is Hauptman's second stint as Seashore superintendent.

Fire Island hit by hurricane.

Comparison of Dune Crestline Position, Psuty.

Archives of the Fire Island National Seashore: William Floyd Estate Registers of Information for Archives and Manuscript Collections. CRBIB Number: 401159. Holdings: NESO.

1992: Fire Island hit by most severe hurricane since 1962. Approximately 100 private homes damaged or destroyed.

Hurricane reshapes the shore and overwhelmed natural sand accretion and caused a second tidal inlet to develop known as Little Pike's Inlet.

Community of Fire Island Pines begins string of self-financed beach renourishing projects.

"Vehicle Free Areas" and "Pedestrian Escort Zones" developed and implemented to study the Piping plover.

Graduate research on nesting and feeding requirements of Piping Plover started in 1992.

Dune crest photography completed, but rendered useless by December Nor'easter.

1993: Distribution of Certificates of Suspension of Authority for Acquisition by Condemnation begins for qualified applicants.

December: Old boat house moved to the Patchogue Maintenance Facility.

Environmental protection of the William Floyd Estate upgraded by installation of new oil and gas furnaces in the main house.

September 30: All use and occupancy rights on twenty-one federal properties in Otis Pike High Dune Wilderness Area expires, leaving only one life estate.

Funding received for planning and design of a Watch Hill ferry terminal center, to be developed in conjunction with the Village of Patchogue River Redevelopment Plan. Plans for the visitor center also included an Education Center in cooperation with the Sea Grant Program of Stony Brook University.

YCC crew constructs helicopter landing pad at Watch Hill to service the Watch Hill area and neighboring community of Davis Park

Recreational visitation to Seashore administered areas reaches peak of 913,875.

A five-year deer immunocontraception research program, sponsored by the Humane Society and in which the park continues to cooperate, begins on the western end of Fire Island.

Graduate research on nesting and feeding requirements of these shorebirds started in 1992 completed.

Fire Island hit by March blizzard.

Publication of Esther Newton's *Cherry Grove, Fire Island: Sixty Years in America's First Gay and Lesbian Town* (Boston: Beacon Press, 1993).

July 11: *New York Times Magazine* runs feature by Elizabeth Hawes, "Life on the Edge: How Long Can Fire Island Last?"

In 1993 the Fire Island Pine Property Owner's Association celebrated its fortieth anniversary with a party featuring a fifteen piece orchestra and an appearance by New York City Mayor David Dinkins.

The USCE, with state and county support, close Little Pike's Peak by installing 1,800 linear feet of buried steel sheet pile and pumping of 1.5 million cubic yards of sand onto the breach.

June: USCE begins start of another reformulation study with Initial Project Study Plan.

The New York Coastal Partnership (NYCP) was founded in 1993 by a group of residents who conduct or manage local businesses.

Hurricanes & Hurricane Preparation Report.

Harpers Ferry Center, Division of Wayside Exhibits completes Wayside Exhibit Plan for the park.

1994: USCE begins studies for a comprehensive reformulation project initiated in accordance with an Initial Project Study Plan of prior year.

New York State Task Force on Coastal Resources backs away from earlier hardline position, adopting a policy of "retreat and selective protection." This approach would foster decisions that protect the public's interest and respect private property rights, while working within the realities of natural coastal processes."

Privately funded sand renourishment projects at many of the developed communities completed.

1995: Memorandum of Agreement signed with Friends of the William Floyd Estate for that group to raise funds to improve visitor facilities at the estate.

Installation of storm windows, a physical security system, an upgraded electrical panel, and an upgraded intrusion and fire systems completed at William Floyd Estate

William Floyd Cultural Landscape Assessment completed for U.S. Department of Interior, National Park Service,

Fire Island National Seashore. Cultural Landscape Assessment of William Floyd Estate completed, as was the first phase of construction for employee housing at the William Floyd Estate. CRBIB Number 403894. Holdings: FIIS.

Marked drop in visitation to Seashore from 810,483 in 1994 to 436,067 in 1995. Attributed to storm events of early 1990s and change in counting techniques.

Fire Island National Seashore, William Floyd Estate: Collections Emergency Operations Plan. CRBIB Number: 401340. Holdings: NESO.

First phase of construction for employee housing and a curatorial storage facility at William Floyd Estate completed

1996: Completion of *Fire Island National Seashore Draft Statement for Management: 1996*.

Amendments made to *Wilderness Area Management Plan* of 1984.

December: Fire Island Lighthouse Preservation Society signs an agreement with the National Park Service to take over the maintenance and operation of the Fire Island Lighthouse and Keeper's Quarters.

John Hauptman leaves post of superintendent of Seashore.

Fire Island hit by hurricane.

Staff residential quarters were completed at the William Floyd Estate, at a cost of roughly $250,000.

1997: Constantine Dillon named superintendent of Seashore.

Dillon receives high marks from environmentalists but clashes often with Fire Island residents over many issues.

1998: Barrett Beach donated to Seashore by the Town of Islip.

Seashore's mosquito surveillance and management program implemented in response to public concern over Eastern Equine Encephalitis (EEE).

Superintendent initiates internal review of the regulations governing motor vehicle use on Fire Island National

Seashore lands. This is in response to concerns of local residents, visitors, and Seashore staff, the Superintendent sought to explore the possibility of building consensus on new regulations controlling motor vehicle use within Seashore.

January: *Enironmental Assessment of Endangered Species Habitat Management*, completed. CRBIB Number: 401671. Holdings: BOSO.

Curatorial storage facility at William Floyd Estate was completed, at a cost of roughly $250,000.

1999: Plans for new ferry terminal and visitor contact facility in Patchogue developed.

Mosquito surveillance and monitoring program expanded in response to discovery of West Nile Virus in New York metropolitan area.

Fire Island hit by hurricane.

2000: The Friends of Fire Island National Seashore, a nonprofit organization established to advocate for Seashore.

Evidence of West Nile Virus-infected mosquitos detected in the community of Saltaire.

Working draft of *Strategic Plan for Fire Island National Seashore: 2001–2005* prepared and posted on Seashore website.

2001: Evidence of West Nile Virus-infected mosquitos detected at Watch Hill.

Fire Island National Seashore Waterborne Transportation System Plan, U.S. Department of Transportation, completed.

2002: Talisman restroom and picnic area improved.

West Nile Virus again detected at Watch Hill and in Otis Pike High Dune Wilderness Area.

Mosquito Surveillance and Management Protocol: Fire Island National Seashore—2002 and *Mosquito Action Plan* completed and posted to Seashore website.

Personal watercraft prohibited in all National Park Service waters, including Fire Island National Seashore.

2003: Completion of *Negotiated Rulemaking Advisory Committee for Off-Road Driving Regulations at Fire Island National Seashore: Final Consensus Agreement.*

Publication of first issue of *Soundings*, an annual report on the Seashore produced by Seashore staff.

The Friends of Fire Island National Seashore see one of its projects materialize with the installation of the Old Inlet Comfort Station, an environmentally friendly, solar-powered composting toilet.

New concession and maintenance building completed at Talisman.

Seashore reaches twenty breeding pairs of Piping Plovers, fifty-two breeding pairs of Least terns, and two breeding pairs of Roseate terns. Thirty-five plover hatchlings and one Least tern hatchling successfully fledged.

West Nile Virus isolated from William Floyd Estate mosquitos. First West Nile Virus-positive collected in the community of Cherry Grove.

May: Constantine Dillon replaced by superintendent of Seashore, David Spirtes.

2004: David Spirtes dies. Replaced by Acting Superintendent Barry Sullivan. Later replaced by Acting Superintendent Dave Reynolds.

New regulations are scheduled to go into effect that will allow personal watercraft in some Seashore waters.

Michael T. Reynolds, Chief of Planning for Yosemite National Park, is named superintendent.

Notes

Chapter 1: Fire Island—Historical Background

1. Madeleine C. Johnson, *Fire Island: 1650s–1980s* (Mountainside, NJ: Shoreland Press, 1983).

2. *Records. Town of Brookhaven up to 1800 as Compiled by the Town Clerk* (Patchogue: Printed at the Office of *The Advance*, 1880), 194; William Wallace Tooker, *The Indian Place-Names on Long Island and Island Adjacent with their Probable Significations* (New York: G. P. Putnam's Sons, *The Knickerbocker Press,* 1911), 243; Johnson, *Fire Island*, pp. 1–2 ; Lavern A. Wittlock, Sr., "Conjecture on Origin of 'Fire Island,' " *Long Island Forum* (September 1973): 165.

3. Richard M. Bayles, *Historical and Descriptive Sketches of Suffolk County and its Towns, Villages, Hamlets, Scenery, Institutions of Important Enterprises with a Historical Outline of Long Island from its First Settlement by Europeans* (Port Jefferson, New York: published by the author, 1874); Martha Bockee Flint, *Early Long Island: A Colonial Study* (New York: G. P. Putnam's Sons, 1896), 70; George L. Weeks, Jr., *Isle of Shells* (New York: Buys Bros., 1965), 163; Thomas R. Bayles, *Early Years in Brookhaven Town* (Brookhaven, New York: Bicentennial Commission, 1975), 7; Johnson, *Fire Island*, p. 2.

4. Osborne Shaw, Letter to the Editor, *The Advance*, November 22, 1956; August Stout, Jr., *Pictorial Memories of the Moriches Area* (New York: n.p. 1973), 83.

5. United States Department of the Interior, *Draft Environmental Impact Statement, Master Plan for Fire Island* (Patchogue, New York: National Park Service, Fire Island National Seashore, 1975): 202–203; Johnson, *Fire Island*, pp. 15, 16.

6. Johnson, *Fire Island*, pp. 26, 27.

7. Ibid., 47–60; Tom Morris, "Fire Island: From Pirates to Slavery to Fun in the Sun," *Newsday* (accessed 23 October 2003, available at http://www.newsday.com/extras/lihistory/spectown/hist0071.htm).

8. Chester R. Blakelock, *History of the Long Island State Parks* (New York: Long Island Forum, 1959), 3; Johnson, *Fire Island*, p. 50; Morris, ". . . From Pirates to Slavery."

9. Johnson, *Fire Island*, pp. 61–77; Morris, "... From Pirates to Slavery"; Bill Bleyer, "A Lifeline for Sailors," *Newsday* (accessed on 23 October, 2003), available at http://www.newsday.com/extras/lihistory/5/hs520a.htm).

10. Johnson, *Fire Island*, pp. 149–160; Randall J. LeBoeuf, Jr., "The Sunken Forest," *Long Island Forum* 34 (April 1971): 81.

11. Richard M. Bayles, *Historical Sketches*, p. 31.

12. Carl A. Starace, "Fire Island Surf Hotel," *Long Island Forum* 33 (August 1970): 156–159; Johnson, *Fire Island*, pp. 32–40; Blakelock, *Long Island State Parks*, pp. 3–4.

13. Johnson, *Fire Island*, pp. 91–97; George DeWan, "The 1892 Cholera Panic," *Newsday* (accessed on October 23, 2003, available at http://www.newsday.com/extras/lihistory/histpast/past11.htm).

14. *General Management Plan*, Fire Island National Seashore (Patchogue, New York: n.p. 1977), 10; James A. Revson, "Fire Island: An Island of Extremes," *Newsday,* July 11, 1985, part 3, 8, 9.

15. Johnson, *Fire Island*, pp. 112–117; Point O'Woods Web Site (accessed on October 31, 2003, available at http://www.fairharbor.com/pl_pointowoods.htm).

16. Ibid.

17. Johnson, *Fire Island*, p. 118.

18. Ibid., 119–120; Lonelyville Web Site (accessed on October 31, 2003, available at http://www.fairharbor.com/pl_lonelyville.htm).

19. Elizabeth Hawes, "Castles in the Sand," *New York Times Magazine*, (July 11, 1993), 29.

20. Johnson, *Fire Island*, pp. 120–126; Ocean Beach Web Site (accessed on October 29, 2003, available at http://www.ocean-beach.com/pl_ob.htm); Andrea Tuppin, "Staying On," *New York Times,* February 14, 1999, Long Island Sunday Supplement.

21. Johnson, *Fire Island*, pp. 142–143.

22. Ibid., 126–129; Mason J. Tufton, "Early History of Saltaire," in Saltaire Citizen's Advisory Association, 20; Saltaire Web Site (accessed on October 31, 2003, available at http://www.fairharbor.com/sa/pl_saltaire.htm).

23. Esther Newton, *Cherry Grove, Fire Island: Sixty Years in America's First Gay and Lesbian Town* (Boston: Beacon Press, 1993), 11, 28; Clifton Hood, "New Studies in Gay and Lesbian History," *Journal of Urban History* (September 24, 1998): 790; Johnson, *Fire Island*, p. 130.

24. Johnson, *Fire Island*, pp. 131–132

25. Ibid., 132–133.

26. Ibid., 133–134; Fair Harbor Web Site (accessed on October 31, 2003, available at http://www.fairharbor.com).

27. Johnson, *Fire Island*, pp. 134–136; Elizabeth Hawes, "Life on the Edge: How Long Can Fire Island Last?," *New York Times Magazine*, (July 11, 1993), Ocean Bay Park Web Site (accessed on October 31, 2003, available at http://www.fairharbor.com/pl_obp.htm).

28. Johnson, *Fire Island*, pp. 136–137.

29. Ibid., 137–138; *Fire Island Guide* (Ocean Beach, New York: Trigor Publishing, 1980), 62.

30. Johnson, *Fire Island*, p. 139; Dunewood Web Site (accessed on November 3, 2003, available at http://www.fairharbor.com/pl_dunewood.htm).

31. Johnson, *Fire Island*, pp. 140–142; Beth Whitehouse, "Shack to Chic on Fire Island," *Newsday,* July 7, 1993, 6.

32. Johnson, *Fire Island*, p. 146.

33. Ibid., 148.

Chapter 2: Fire Island Erosion: Shifting Sands and Shoreline of a Barrier Island

1. S. P. Leatherman and J. R. Allen, eds., *Fire Island Inlet to Montauk Point Long Island, New York: Reformulation Study* (Patchogue, NY: United States Department of the Interior, National Park Service, Fire Island National Seashore, 1985), executive summary; Neal Bullington, and David Griese, *Exploring the Sunken Forest*, Fire Island National Seashore, (Patchogue, NY: United States Department of the Interior, National Park Service, Fire Island National Seashore, 1977).

2. Dr. Henry Bokuniewicz, pers. comm., Marine Sciences Research Center, Stony Brook University.

3. Christopher Schuberth, "Barrier Beaches of Eastern America," *Natural History* (June–July 1970), 53; N. E. Taney, *Geomorphology of the South Shore of Long Island, New York* (U.S. Army Coastal Eng. Research Center, Tech. Memo, 128), 50; Kenneth Ruzyla, "Effects of Erosion on Barrier-Island Morphology Fire Island, New York," in *Coastal Morphology*, ed. Donald R. Coates (Binghamton, NY, State University of New York, 1972), 220.

4. *Hurricane Damage Mitigation Plan for the South Shore Nassau and Suffolk Counties, N.Y.* (Hauppauge, New York: Long Island Regional Planning Board, 1984); Albert Ruhfel, "The Fire Island Inlet," *Long Island Forum* 34 (1971): 90–93; Ruzyla, "Effects of Erosion," p. 234.

5. Madeleine C. Johnson, *Fire Island: 1650s–1980s* (Mountainside, NJ: Shoreland Press, 1983), 11; Ruhfel, "Effects of Erosion," p. 92.

6. F. L. Panuzio, *The Atlantic Coast of Long Island*, Proc. 11th Conference of Coastal Engineers, ed. J. W. Johnson (Richmond, CA: Council on Wave Resistence, 1968); Warren E. Yasso and Elliott M. Hartman, Jr., *Beach Forms and Coastal Processes*, MESA New York Bight Atlas Monograph 11 (New York: New York Sea Grant Institute, 1975), 21; Leatherman and Allen, *Fire Island Inlet*, p. iii.

7. Leatherman and Allen, *Fire Island Inlet*, p. iii; Jay Tanski and Henry Bokuniewicz, *Special Study No. 144* (Stony Brook, NY: New York Sea Grant Institute, 1992); Ruzyla, "Effects of Erosion," p. 235; *Proposed Long Island South Shore Hazard Management Program* (Hauppauge, NY: Long Island Regional Planning Board, 1989), 2.3.

8. Ruzyla, "Effects of Erosion," p. 222; Kenneth Ruzyla, "Effects of Erosion on Barrier-Island Morphology, Fire Island, New York," in *Barrier Beach Dynamics* (Hauppauge, NY: Suffolk County Planning Department, n.d.); Thore Omholt, "Coastal Processes and Shore Protection: Address to the Marine Resources Council Committee on Beach Erosion," in *Barrier Beach Dynamics* Hauppauge, NY: Suffolk County Planning Department, n.d.).

9. Johnson, *Fire Island*, p. 173

10. Ernest S. Clowes, *The Hurricane of 1938 on Eastern Long Island*, (Bridgehampton, NY: Hampton Press, 1939), 9–10.

11. Ibid., 22–23

12. Paul C. Focazio and Jay Tanski, "Monitoring Change," *Coastlines* 29 (Winter 2000) (accessed on January 26, 2003. Available at http://www.seagrant.sunysb.edu/Pages/Coastlines/Winter00.htm#part2).

13. Johnson, *Fire Island*, p. 173; Daniel Conley, "The Short History of Little Pike's Inlet," *Coastlines* 29 (Winter 2000) (accessed on January 26, 2003. Available at http://www.seagrant.sunysb.edu/Pages/Coastlines/Winter00.htm#part2); William Daley and Clifford Jones, "The Westhampton Dunes Story: A Blueprint for Coastal Management," *Shore and Beach* 68 (January 2000): 2.

14. *Guidelines for Long Island Coastal Management* (Hauppauge, NY: Long Island Regional Planning Board, 1973), 5.

15. T. Saville, Jr., *Sand Transfer, Beach Control, and Inlet Improvements, Fire Island Inlet to Jones Beach*, ed. J. W. Johnson (Richmond, CA: Proceedings of the 7th Conference of Coastal Engineers, 1961), 785–807.

16. Yasso and Hartman, Jr., *Beach Forms*, 22.

17. *Fire Island to Montauk Point, New York, combined Beach Erosion Control and Hurricane Protection Project* (Place ????: United States Army Corps of Engineers, 1959).

18. "$37 Million Erosion Plan Passes 1st Major Hurdle," *Newsday*, October 27, 1959.

19. "18,000 Xmas Trees To Fight Erosion," *Newsday*, January 7, 1960.

20. *Proposed Long Island South Shore Hazard Management Program*, chap. 3–22.

21. *Proposed Long Island South Shore Hazard Management Program*, chap. 3–24, 25.

22. "$2,300,000 Erosion Job Gets Underway May 1," *Long Island Press*, January 2, 1957.

23. "Erosion Job Cost Set at $62 million," *Newsday*, October 23, 1957.

24. Byron Porterfield, "Fight to Save L.I. Beaches Will Start Tomorrow," *New York Times*, June 4, 1959; and "House Votes 220 Gs for Fire Island Anti-Erosion Project," *Newsday*, June 3, 1959.

25. "Stuy Hits Moses Again on Inlet Project," *Newsday*, April 11, 1960.

26. "Fire Island Dike Called a Danger to Small Boats," *Long Island Press*, January 14, 1960.

27. "Top Army Engineer Calls Fire Island Project a Success," *Long Island Press*, May 12, 1960.

28. "Propose New Inlet at Fire Island," *Long Island Press*, March 15, 1960.

29. "Proposed Fire Island Inlet May Cause Havoc in Bay," *Long Island Press*, March 21, 1960.

30. Tom Collins, "State Discloses New Plans for Fire Island Inlet Channel," *Newsday*, January 11, 1960.

31. "Red Tape Not Halting Fire Island Erosion," *Newsday*, March 5, 1962.

32. Bare $36 Million Plan to Control L.I. Erosion," *Newsday*, May 9, 1959; Tom Renner, "Supervisors Cool to Army Erosion Plan," *Newsday*, May 12, 1959.

33. Walt Brevig, "Liers to Ask What's Score on Erosion," *Newsday*, August 11, 1959.

34. "Set Dedication of Forest Preserve," *Newsday*, August 12, 1959.

35. "Erosion Control Project May Begin Within 2 Months," *Long Island Press*, August 16, 1963.

36. "Suffolk Stalls Erosion Job Verdict," *Newsday*, August 6, 1963.

37. "Fire Island Inlet to Montauk: Fact Sheet" (New York: United States Army Corps. of Engineers, 2003); *Fire Island to Montauk Point Reformulation Study: Vision Statement* (New York: U.S. Army Corps of Engineers, Power Point Presentation, 2003).

38. Ibid.

39. Ibid., Daley and Jones, "Westhampton Dune Story."

40. Letter of Correspondence between Scott M. Cullen of Nature Conservancy to Bob Spenser, October 24, 2003.

41. Department of Interior Comments to United States Army Corps. of Engineers Draft Environmental Impact Statement, February 29, 2000; Response of the Fire Island Association to the Department of Interior's "Development Moratorium" Comment on the United States Army Corps of Engineer's Draft Decision Document and Draft Environmental Impact Statement for the Fire Island Interim Plan" (Silver Spring, MD: Linowes and Blocher, LLP, March 24, 2000).

42. *Proposed Long Island South Shore Hazard Management Program.*

43. Ibid.; *Proceedings of the First Long Island Coastal Alliance Conference,* 1990.

44. *Now and for the Future: A Vision for New York's Coast: Recommendations of the Governor's Task Force on Coastal Resources* (November 1991).

45. Spencer and Terchunian, "The Sand Thieves of Long Island's South Shore," *Shore and Beach* (July 1997): 8.

46. "The Fire Island Interim Project Will Not Cause Building on Dunes," Fire Island Association White Paper (April 2000).

47. Interview with Gerald Stoddard of Fire Island Association, December 22, 2003.

48. Johnson, *Fire Island,* pp. 6–7.

49. Lee F. Koppelman, *Economic Analysis—Fire Island Reach 1995* (Hauppauge, NY: New York Coastal Partnership), 43.

Chapter 3: The Creation of the Fire Island National Seashore

1. Barry Mackintosh, *The National Park Service: A Brief History* (Washington, DC: National Park Service, 1999). Accessed on November 19, 2003 available at http://www.cr.nps.gov/history/hisnps/NPSHistory/npshisto.htm

2. Ibid.

3. Ibid.

4. Mackintosh, *Brief History*; Barry Mackintosh, *The National Parks System: Shaping the System,* 5th ed. (Washington, D.C.: National Parks Service,

1985). Accessed on November 19, 2003 available at http://www.cr.nps.gov/history/online_books/mackintosh1/sts2b.htm.

5. Mackintosh, *Shaping the System.*

6. Mackintosh, *Brief History*; Mackintosh, *Shaping the System.*

7. Ibid.

8. Mackintosh, *Brief History*; Mackintosh, *Shaping the System.*

9. Mackintosh, *Shaping the System.*

10. Ibid.

11. Fire Island National Seashore: A Proposal (Washington, DC: United States Department of the Interior, National Parks Service, 1964).

12. Mackintosh, *Shaping the System*

13. Ibid.

14. Ibid.

15. Ibid.

16. Mackintosh, *Brief History.*

17. Johnson, *Fire Island*, pp. 91–97; George DeWan, "The 1892 Cholera Panic," *Newsday* (accessed on October 23, 2003, available at http://www.newsday.com/extras/history/histpast/past11.htm)

18. William Norris Leonard and Lee E. Koppelman, *People and Parks*, Suffolk County Planning Commission, November 1960; and Lee E. Koppelman, *Planning for Open Space in Suffolk County*, Comprehensive Plan Series, volume II, Suffolk County Planning Department, May 1964.

19. *The Race for Open Space* (New York, NY, The Regional Plan Association of New York, 1960).

20. "The Moses Plan," *Long Island Press*, March 25, 1962.

21. "Moses' Fire Island Plan Okayed By State Shoreline Commission," *Smithtown Star*, August 1, 1962; "Moses Offers Plan to Prevent New Storm Devastation on the Atlantic Shore," *Long Island Press*, March 26, 1962, 25, 36.

22. *Suffolk County News*, July 27, 1956. Representative Stuyvesant Wainwright called for the creation of a National Seashore based on the National Park Service Report, *Our Vanishing Shoreline*. The hearing held on August 22, 1957 echoed this recommendation as a counter strategy against Moses' road proposal.

23. Temporary State Commission on Protection and Preservation of the Atlantic Shore Front, Joseph F. Carlino, Chairman.

24. "Hearing Scheduled for July 10," *News Review*, June 10, 1962. The Commission included among the fifteen members County Executive Dennison, Assembly Speaker Duryea, Presiding Supervisor Arthur Cromarty, and State Senator Elisha T. Barrett. "State Erosion Unit Details Plan for L.I.," *Newsday*, July 25, 1962. The thirty-six page report supported the Moses road proposal. "OK Due Today on F.I. Road Plan," *Newsday*, July 26, 1962.

25. Vicky Panner, "Javits, Pike Open Hearing on Fire Island Shore Bill," *L.I. Commercial Review*, October 1, 1963. In addition, hearings were held before the Subcommittee on National Parks of the Committee on Interior and Insular Affairs of the House of Representatives, Eight-Eighth Congress on H.R. 3693, 4999, 6111, 6934, 6936, 7107, 7297, 7359, 7512; "Bills to Establish the Fire Island National Seashore in the State of New York," September 30, 1963 at Islip, N.Y., and April 10, 1964 at Washington DC; Wayne N. Aspinall, Colorado, Chairman. Paul B. Townsend Editor and Publisher, "Fire Island Hear-

ings: Favor 33 Mile National Seashore," *L.I. Commercial Review*. Udall cut back from 52 miles to 33 miles—no road, no additional access.

26. Ronald Maiorana, "Park Bill to Pass Fire Island Told," *New York Times*, October 1, 1963. Representative Otis Pke said eleven bills had been introduced in the House and Senate.

27. "Suffolk Votes to Buy Tract on Fire Island," *Long Island Press*, August 13, 1963. On August 12, 1963 the Suffolk County Board of Supervisors followed the Planning Commission's recommendation and approved the acquisition of the four-mile stretch between the eastern boundary of the Smith Point County Park and the Moriches Inlet County Park. This gave the County control over twelve miles or more than one-third the length of Fire Island. The State (Moses) provided 75 percent of the cost.

28. Conversation between the Suffolk County Planning Board Director and Jack Luchsinger, Chairman of the Town of Brookhaven Planning Board and the Brookhaven representative on the Suffolk County Planning Commission [1962]; and, Lee E. Koppelman, *Report on the Need and Feasibility for County Park Facilities in Suffolk County*, January 1962; and, Lee E. Koppelman, *Planning for Open Space in Suffolk County* (Suffolk County: Suffolk County Planning Department, May 1964), p. 93.

29. "Rocky Says State Has No Funds for Fire Island Road," *L.I. Commercial Review*, February 27, 1963, p.1; "New Snag Perils Erosion Control Plans," *Newsday*, February 26, 1963. Rockefeller kills Moses road plan saying "no funds"; "Seems Fire Island Road's Almost Washed Up! Barbash Happy as Moses Leaves the Long Island State Park Commission," *New York Standard*, January 14, 1963.

30. Public Law 88-587, 88th Congress, S. 1365, September 11, 1964.

31. *Our Vanishing Shoreline—Seashore Recreation Area Survey of Atlantic and Gulf Coasts* (Washington, DC: Department of the Interior, National Park Service, 1963); Virginia Bennett Moore, "A Possible National Seashore," *National Parks Magazine*, February 1963.

32. U.S. House, Committee on Interior and Insular Affairs H.R. 3693, February 11, 1963.

33. *The Race for Open Space* (New York, NY: The Regional Plan Association of New York, 1960).

34. William F. Leonard and Lee E. Koppelman, *People and Parks* (Hauppauge, NY: n.p., November 1960).

35. "Udall Reaffirms Support for Fire Island Seashore,"*Long Island Commercial Review*, April 2, 1963.

36. Ibid.

37. Correspondence of April 3, 1963 from L. J. Pennachio to Supt. of Fire Island National Seashore.

38. U.S. House, Committee on Interior and Insular Affairs H.R. 4999, March 19, 1963.

39. U.S. Senate, Committee on Interior and Insular Affairs S. 1365, April 25, 1963.

40. Letters from: Charles Collingwood to Senator Javits, April 26, 1963; Gardner J. Stout (Wall Street), April 29, 1963; Clifford W. Michel (Lock, Rhoades & Co.), June 12, 1963. In opposition to Udall: Frank W. Wyman (Park Avenue),

Leonard B. Frutkin (5th Avenue) Secretary of the Dune Road Property Owners of Westhampton Beach.

41. "Suffolk Board to Back Fire Island Park," *Newsday*, April 3, 1963.

42. "Supervisors Back Fire Island Park, Delay Erosion Plan," *Long Island Press*, April 9, 1963.

43. U.S. House, Committee on Interior and Insular Affairs H.R. 6111, Rep. James R. Grover, May 7, 1963; U.S. House, Committee on Interior and Insular Affairs H.R. 6213, Rep. John Wydler, May 9, 1963; U.S. House, Committee on Interior and Insular Affairs H.R. 6934, Rep. Leon W. O'Brien, June 11, 1963; U.S. House, Committee on Interior and Insular Affairs H.R. 6936, Rep. William Fitts Ryan, June 11, 1963; U.S. House, Committee on Interior and Insular Affairs H.R. 7297, Rep. Edna F. Kelly, June 26, 1963; U.S. House, Committee on Interior and Insular Affairs H.R. 7359, Rep. Eugene J. Keogh, July 1, 1963; U.S. House, Committee on Interior and Insular Affairs H.R. 7107, Rep. Otis Pike, June 18, 1963.

44. Leonard Baker, "Pike Shifts, Backs U.S. Park as Only Hope for Fire Island," *Newsday*, April 10, 1963.

45. U.S. House, Committee on Interior and Insular Affairs H.R. 7107, Representative Otis Pike, June 18, 1963.

46. This eight mile zone was recommended by the NPS as meeting the criteria for designation in accord with the Wilderness Act of 1964 requirement (PL 88-55). U.S. Senate, Committee on Interior and Insular Affairs S.304 was introduced by Senators Javits and Moynihan to designate the "Otis Pike-Fire Island High Dune Wilderness," on August 19, 1980; and companion U.S. House, Committee on Interior and Insular Affairs H.R. 7814 introduced by Representatives William Carney and Norman Lent.

47. Albin Krebs, "Fire Island Seashore–Udall's Latest Plan," *The New York Herald Tribune*, April 9, 1963.

48. "Governor Avoids Highway Approval," Ocean Beach, NY, *Fire Island News*, July 21, 1962, p.1

49. Letter and attached memorandum from Moses as Chairman of the Triborough Bridge and Tunnel Authority dated October 24, 1963 to Senator Javits.

50. Ibid.

51. William Barber, "Moses' Pipe Dream? The Master Planner Would Pave Fire Island," *Newsday*, April 4, 1974.

52. Personal correspondence from Moses to Godbolt, March 8, 1974.

53. William Barber, "Moses' Pipe Dream? The Master Planner Would Pave Fire Island," *Long Island Press*, April 7, 1974.

54. Personal correspondence from Senator Jackson to Senator Javits. The delay in settling the hearing was occasioned by the need of the Committee to receive reports on the matter from the NPS.

55. Leonard Baker, "Say Shore Plan Grew From a Moses Acorn," *Newsday*, June 1963.

56. Leonard and Koppelman, *People and Parks*.

57. Lee E. Koppelman, *Comprehensive Open Space Plan*, May 1964.

58. Conversation at the Weld home between David Weld and the Suffolk County Planning Director, May 1964.

59. U.S. House, Committee on Interior and Insular Affairs H.R. 8035 approved October 9, 1965.

60. U.S. Senate, Committee on Interior and Insular Affairs S.2148—Javits and Kennedy.

61. U.S. House, Committee on Interior and Insular Affairs H.R. 8035 became Public Law 89–244, October 9, 1965. Senator Bible, Chairman of the Senate Committee on Interior and Insular Affairs submitted Report No. 748 in support of the House bill.

Chapter 4: Historical and Cultural Resources

1. Fire Island National Seashore Draft Statement for Management, 1979, p. 4; Larraine A. Fletcher and Ellen R. Kintz et. al., *Fire Island National Seashore Cultural Resources Inventory* Division of Cultural Resources (Boston, MA: U.S. Department of Interior, National Park Service, 1979); John F. Vetter and Bert Salwen, *Report of an Archeological Reconnaisance of Fire Island, Suffolk County, New York* (Patchogue, NY: U.S. Department of Interior, National Park Service, Fire Island National Seashore, 1979).

2. *Draft Master Plan*, 10, 18–19; Collin Nash, "Focus on History: William Floyd Estate," *Newsday*, July 2, 1995, 76; *Fire Island Seashore Statement for Management* (Patchogue, NY: U.S. Department of Interior, National Park Service, Fire Island National Seashore, 1979), 4; *Fire Island National Seashore Draft Statement for Management, 1996*, p. 9.

3. Phone interview with Steve Kesselman, from Hauppauge to Salem Maritime, November 3, 2004.

4. *Fire Island National Seashore Draft Statement for Management, 1996*, p. 16.

5. *Fire Island National Seashore Draft Statement for Management, 1996*, p. 17; *Description of Selected Alternatives Site Development and Facilities: William Floyd Estate, Fire Island National Seashore* (Washington, DC: U.S. Department of Interior, National Park Service, Fire Island National Seashore, 1979); *William Floyd Estate Cultural Landscape Assessment: Draft* (Patchogue, NY: U.S. Department of Interior, National Park Service, Fire Island National Seashore, 1995).

6. This and a portion of what follows is provided by the Fire Island Preservation Society in various promotional materials.

7. *Annual Narrative Report of Superintendents: Fire Island National Seashore* (Patchogue, NY: n.p. 1987).

8. Robert G. Muller, *Long Island's Lighthouses Past and Present* (Patchogue, NY: Long Island chapter of the US Lighthouse Society, 2004), 314; Bill Bleyer, "Bringing History to Light," *Newsday*, April 15, 2003, A18.

9. Thompson, *Administrative History*, p. 10; *Fire Island National Seashore Draft Statement for Management,* 1996, p. 26; *Soundings: 2003 Annual Report*, 7.

10. *Fire Island National Seashore Draft Statement for Management, 1996*, p. 14.

11. Ibid., 7–8.

12. Thompson, *Administrative History*, p. 11; *Fire Island National Seashore Draft Statement for Management, 1996*, p. 29.

13. Ibid., 11; Eden Lipson, "When the Dancing Had to Stop," *New York Times*, September 5, 1976.

14. *Annual Narrative Report of Superintendents: Fire Island National Seashore* (Patchogue, NY: n.p. 1994).

15. Thompson, *Administrative History*, pp. 11–12; *Fire Island National Seashore Draft Statement for Management, 1996*, p. 26; *Soundings, 2003 Annual Report*, 7.

16. Thompson, *Administrative History*, p. 13; *Fire Island National Seashore Draft Statement for Management, 1996*, p. 27; *Soundings: 2003 Annual Report*, 8.

17. Thompson, *Administrative History*, p. 14; *Fire Island National Seashore Draft Statement for Managment, 1996*, p. 28.

18. Ibid.

19. Data assembled from various *Annual Narrative Report of Superintendents: Fire Island National Seashore*

20. *Environmental Assessment for Endangered Species Habitat Management* (Patchogue, NY: U.S. Department of Interior, National Park Service, Fire Island National Seashore, 1998).

21. *Environmental Assessment for Endangered Species Habitat Management* (Patchogue, NY: U.S. Department of Interior, National Park Service, Fire Island National Seashore, 1998), 9

22. Ibid., 10

23. Ibid.

24. *Fire Island National Seashore Draft Statement for Management, 1996*, p. 15; *Strategic Plan for Fire Island National Seashore: 2001–2005* (Patchogue, NY: U.S. Department of Interior, National Park Service, Fire Island National Seashore). Available at http://www.nps.gov/fiis/stratplanFY01-05.htm. Accessed on May 5, 2004; *Soundings: 2003 Annual Report*, 5.

25. *Deer and People at Fire Island: How You Can Help Protect the Health of Deer and People* (Patchogue, NY: U.S. Department of Interior, National Park Service, Fire Island National Seashore), undated, accessed at http://www.nps.gov/fiis/deerpeople/deer.html.

Chapter 5: Seashore Management

1. *Fire Island National Seashore: A Proposal*, Northeast Regional Office and Eastern Office of Design and Construction, National Park Service (Washington, DC: United States Department of the Interior, September 1963).

2. *Draft Master Plan* (Patchogue, NY: U.S. Department of Interior, National Park Service, 1975), p.6

3. Marsha L. Hamilton, "Shifting Sand: Long Island's Barrier Beaches," *Long Island Historical Review* 16 (Fall 2003/Spring2004).

4. *Fire Island National Seashore: A Proposal* (Washington, DC: U.S. Department of Interior, National Park Service, U.S. Bureau of Outdoor Recreation, March 1964).

5. Ibid.

6. *The Master Plan: Fire Island National Seashore* (Patchogue, NY: U.S. Department of Interior, National Park Service, Fire Island National Seashore, May 1965).

7. Maurice Swift, "U.S. to Acquire Land, 31 Homes on Fire Isle," *Newsday*.

8. Betty Thompson, *Administrative History of Fire Island National Seashore* (Patchogue, NY: United States Department of Interior, National Park Service, Fire Island National Seashore, 1977).

9. P.L. 88-587 (All) FNS Enabling Legislation, 1964.

10. Public Law 89–244 (All) William Floyd Estate (1965).

11. *Conceptual Master Plan For Fire Island National Seashore* (Washington, DC: United States Department of Interior, National Parks Service, Denver Service Center, Denver, 1973); Thompson, *Administrative History*, p. 5.

12. Robert Fresco, "Fire Island Falls Short of Goals," *Newsday*, September 15, 1974.

13. Ibid.

14. Robert Fresco, "Fire Island Falls Short of Goals," *Newsday*, September 15, 1974.

15. "Preserving Fire Island," *Newsday,* September 17, 1974.

16. Press release from the Office of Senator Javits, February 27, 1975.

17. The United States Court of Appeals for the Second Circuit in Biderman v. Morton, 497, F.2d 1141 (2d cir. 1974).

18. The United States Court of Appeals for the Second Circuit in Biderman v. Morton, 497, F.2d 1141 (2d cir. 1974), 1148–49.

19. Press release from the Office of Senator Javits, November 13, 1975 on his statement before the Senate Interior Subcommittee on S.867, Fire Island National Seashore.

20. The Javits bill was passed on April 7, 1976.

21. *Land Protection Plan* (Patchogue, NY: U.S. Department of Interior, National Park Service, Fire Island National Seashore, 1984), 21.

22. U.S. Congress. *National Parks and Recreation Act of 1978*, H.R. 12536, 94th Congress, *Congressional Record* (July 11, 1978).

23. Letter to Senator Jacob Javits from George Biderman of July 26, 1978.

24. Letter to Representative Thomas Downey, July 20, 1978, from George Biderman, President of the Fire Island Association, Inc.

25. Letter with enclosures to the Regional Director of the North Atlantic Region of the NPS on the justification for the establishment of the Dune District, from Richard W. Marks, Superintendent F.I.N.S.

26. Letter with enclosures to the Regional Director of the North Atlantic Region of the NPS on the justification for the establishment of the Dune District, from Richard W. Marks, Superintendent F.I.N.S.

27. Paul J. Godfrey, *Management Guidelines for Barrier Beaches* (Boston, MA: University of Massachusetts, 19XX), p.4.

28. Richard F. Babcock and Richard J. Roddewig, *Report to the National Park Service on Fire Island National Seashore* (Chicago: Ross, Hardies, O'Keefe, Babcock & Parsons, 1975), 62.

29. Among the strongest writings cited were the following: Warren E. Yasso and Elliott M. Hartman, *Beach Forms and Processes* (NY: New York Sea

Grant Institute, April 1976); Orrin H. Pilkey, Jr., *How to Live with an Island* (Raleigh, NC: North Carolina Department of Natural Resources, 19XX); Lee E. Koppelman, *Land Use, Zoning and Development Control Policies—Fire Island National Seashore* (Hauppauge, NY: Suffolk County Department of Planning, 1975); Ian L. McHarg, *Design with Nature* (Garden City, NY: The Natural History Press, 1969).

30. Letter to Vincent Ficara, April 18, 1978, from Paul X. Colter.

31. Letter July 7, 1978, from Vincent Ficara. Re: Tract 2518.

32. Islip Zoning Ordinance, Article VA, 568–591.

33. *Land Use, Zoning and Development Control Policies—Fire Island National Seashore*: p. 6.

34. *Final Environmental Impact Statement for Fire Island Inlet to Montauk Point, New York* (New York, NY: Beach Erosion Control and Hurricane Protection Project. U.S. Army Engineer District, September 1977), 361.

35. Letter to George Biderman, March 26, 1979, from Superintendent Marks.

36. Letter to Louis J. Pennachio, Davis Park Association, March 27, 1979, from Superintendent Marks.

37. Letter to Marks, March 29, 1979, from Pennachio.

38. Letter to Pennachio, April 2, 1979, from Marks.

39. Letter to Marks, April 30, 1979, from Pennachio.

40. Letter to Pennachio, May 15, 1979, from Marks.

41. Deposition of Richard Marks of March 1979 in U.S.A. v. Vincent Ficara, et al.

42. Correspondence of July 5, 1979 and July 10, 1979 between Senator Ted Stevens and the Director of the NPS.

43. Letter to Chair of the Islip Town Environmental Council, October 4, 1979, from Charles Pulaski; Letter to Chair of the Brookhaven Conservation Advisory Council, November 6, 1979, from Adelaide Flatau; Resolution sent by Eugene T. Dooley, Brookhaven Town Clerk, December 5, 1979; Letter to Deputy Supervisor of the Town of Islip, including the Town Board's resolution of support, December 6, 1979, from Michael A. LoGrande; Letter to Chair, Huntington Conservation Board, December 13, 1979, from Joy S. Squires; Letter to President of the Fire Island Wilderness Committee to Senator Javits, from Joseph Zysman; Editorial support from *The Long Island Advance*, December 27, 1979.

44. James Barron, "Fire Islanders Fight U.S. for Their Property," *New York Times*, July 31, 1981, B2.

45. Mid-Fire Island Preservation Association newsletter of December 10, 1979.

46. Letter to Supervisor Jack Lefkowitz, December 10, 1979, from Walter G. Reich.

47. Letter to Senator Javits, December 19, 1979, from Walter G. Reich.

48. *Environmental Assessment: Amendment to Wilderness Area Management Plan (1984)* (Patchogue, NY: National Park Service, Fire Island National Seashore, 1996); *Annual Narrative Report of Superintendents: Fire Island National Seashore* (Patchogue: NY: n.p., 1994).

49. *Fire Island National Seashore: Preliminary Report on Findings and Recommendations* (Washington, DC: The Conservation Foundation, November 12, 1975); Thompson, *Administrative History*, p. 8.

50. *Fire Island National Seashore: Preliminary Report on Findings and Recommendations*, p. 23.

51. *Preliminary Report on Findings and Recommendations*, p. 26; Robert Fresco, "Lifting a Heavy Hand on Fire Island," *Newsday*, November 13, 1977.

52. Babcock and Roddewig, *Report to National Park Service.*

53. Ibid., 10.

54. Ibid., 21.

55. Title 36, Chapter 1, Parts 28.4 and 28.5.

56. Babcock and Roddewig, *Report to National Park Service*, p. 28.

57. *Draft Master Plan*, p. 87.

58. Babcock and Roddewig, *Report to National Park Service*, p. 22.

59. "Seashore Plan Called All Wet," *Long Island Press*, January 20, 1975; "Master Plan: A Primer," *Long Island Press*, May 18, 1975.

60. Glenn Fowler, "The Fire Island Plan Draws Mostly Cheers," *New York Times*, August 1, 1976.

61. Karl Grossman, "Seashore Plan: Winner in 2d Try," *Long Island Press*, July 28, 1976, 4.

62. "Master Plan: A Primer," 6A.

63. Karl Grossman, "Last Long Island Wilderness?," *Long Island Press,* December 5, 1974.

64. Barbara Blumberg, *Celebrating the Immigrant: An Administrative History of the Statue of Liberty National Monument: 1952–1982* (Washington, DC: U.S. Department of the Interior, National Park Service, 1985), 57.

65. *Draft General Management Plan Summary* (Washington, DC: U.S. Department of Interior, National Park Service, Fire Island National Seashore, June 1976), 5; Thompson, *Administrative History*, p. 5.

66. *Land Protection Plan* (Patchogue, NY: Fire Island National Seashore, NPS, United States Department of the Interior, 1984), p. 21.

67. *Draft General Management Plan Summary,* p. 13.

68. Ibid., 16.

69. Grossman, "Seashore Plan," p. 4.

70. Fire Island National Seashore, Off-Road Vehicle Regulations, 36 CFR 7.20(a).

71. *Draft Management Plan Summary*, p. 4; Robert Fresco, "Lifting a 'Heavy Hand' on Fire Island," *Newsday*, November 13, 1977

72. Fresco, "Lifting a Heavy Hand"; Karl Grossman, "Fire Island Finds Federal Friend," *Long Island Press*, August 23, 1976.

73. *Land Protection Plan* (Patchogue, NY: U.S. Department of Interior, National Park Service, Fire Island National Seashore, 1984), pp. 4–5.

74. Katurah Mackay, "Homes Erode National Seashore," *National Parks* (July 1999): 10.

75. *National Parks and Recreation Act of 1978,* Public Law 95–625, 95th Cong., November 10, 1978.

76. Index No. 78 Civ.

77. Letter to Senator Jacob Javits dated October 17, 1978, from Irene E. Sorse.

78. Memo to Senator Javits, November 3, 1978, from Barbara (Javits staff).

79. Memo to Senator Javits, November 8, 1978, from Beverly Charles.

80. Letter to Beverly Charles, November 8, 1978, from Daniel L. Alterman, Esq.

81. Letter to Daniel L. Alterman, Esq., November 8, 1978, from Senator Javits.

82. Shawn G. Kennedy, "U.S. Destroying Squatter Shacks on Fire Island–Dwellers Hope Court Will Delay Park-Site Action," *New York Times,* November 28, 1978.

83. Carol Lee Ritch, "Fire Island Eviction to Proceed," *Newsday,* November 16, 1978.

84. Letter to Senator Javits, November 30, 1978, from Acting Secretary Robert L. Herbst.

85. United States of America against certain lands located in the County of Suffolk, State of New York, et al., 67c705; 68c838

86. United States of America against certain lands located in the County of Suffolk, State of New York, et al., 67c705; 68c838, 89, 90

87. United States of American, Plaintiff—Appellee against Certain Lands Located in the County of Suffolk, State of New York, et. al., Defendants—Appellants Brief for Appellee, on behalf of Edward R. Korman, United States Attorney for the Eastern District of New York.

88. Memo from Beverly Charles to Senator Javits, December 15, 1978.

89. Barron, "Fire Islanders Fight U.S. for Their Property," *New York Times*, July 31, 1981.

90. Telephone interview with Donald Weir, Hauppauge to Buffalo, November 19 and 24, 2004

91. Barron, "Five Island Fight U.S."

92. Report to the Honorable Daniel P. Moynihan, United States Senate (Washington, DC: United States General Accounting Office, 1981).

93. Fire Island National Seashore, Development Standards 36 CFR Ch. 1 28.12

94. Phone interview with Karl C. Sollers, Hauppauge to Queens, November 3, 2004.

95. *Fire Island National Seashore Draft Statement for Management* (Patchogue, NY: U.S. Department of Interior, National Park Service, Fire Island National Seashore, 1996), 7; Interview with Diane Abell, Landscape Architect, Fire Island National Seashore, May 11, 2004.

96. Carole Paquette, "Fire Island: Accent is on Upgrades, and Just Up," *New York Times*, April 22, 2001, 7.

97. Mackay, "Homes Erode National Seashore," *National Parks Magazine* (July 1999): 11.

98. Ibid., 23; Interview with Diane Abell, Landscape Architect, Fire Island National Seashore, May 17, 2004.

99. Thompson, *Administrative History*, p. 10.

100. *Final Transportation Study* (Patchogue, NY: U.S. Department of Interior, National Park Service, Fire Island National Seashore, 1977), II–4.

101. *Draft Master Plan*, 16–18.

102. *A Study of Visitor Use Satisfactions* (New York, NY: U.S. Department of Interior, National Parks Service, Fire Island National Seashore, October 1969).

103. *A Study of Visitor Use Satisfactions* (New York, NY: U.S. Department of Interior, National Parks Service, Fire Island National Seashore, October 1969).

104. *Fire Island National Seashore: Preliminary Transportation Study, Work Directive No. 4–0070–74–4* (Place ?????: Vollmer Associates, July 26, 1974), 5.

105. *Final Transportation Study* (Patchogue, NY: U.S. Department of Interior, National Park Service, Fire Island National Seashore, 1977), 5.

106. *Fire Island National Seashore Waterborne Transporation System Plan* (Patchogue, NY: U.S. Department of Interior, National Park Service, Fire Island National Seashore, February 2001), xi–xii.

107. Ibid., 10.

108. *Soundings: 2003 Annual Report*, p. 12; *Mosquito Surveillance and Management Protocol: Fire Island National Seashore—2002* and *Mosquito Action Plan* (Patchogue, NY: Fire Island National Seashore Mosquito Action Plan Committee, U.S. Department of Interior, National Park Service, Fire Island National Seashore, 2002), available at http://www.nps.gov/fiis/MSMPfy02.htm. Accessed on May 5, 2004.

109. Thompson, *Administrative History*, p. 8.

110. Robert Fresco, "Lifting a 'Heavy Hand' on Fire Island," *Newsday*, November 13, 1977.

111. Walter Reich, Remarks at the Fire Island National Seashore 40th Anniversary Celebration, September 11, 2004

112. *Fire Island National Seashore Draft Statement for Management*, 1996, p. 8.

113. Negotiated Rulemaking Advisory Committee for Off-Road Driving Regulations at Fire Island National Seashore: Final Consensus Agreement (Patchogue, NY: U.S. Department of Interior, National Park Service, Fire Island National Seashore, August 22, 2003); Beth Whitehouse, "Fire I. Fights Over Coveted Permits," *Newsday*, March 6, 1994, 6; *Conflict Assessment: The Prospects for Building Consensus on Fire Island National Seashore's Vehicle Use Regulations: A Report to the Superintendent of the Fire Island National Seashore* (Cambridge, MA: Prepared by Consensus Building Institute, September 1999).

114. Ibid.

115. "National Park Service Moves Forward With Management of Personal Watercraft Use" (Washington, DC: U.S. Department of Interior, National Park Service, April 16, 2002); *Soundings: 2003 Annual Report*, p. 17.

116. Thompson, *Administrative History*, p. 6.

117. *Organization Review* (Patchogue, NY: United States Department of Interior, National Park Service, Fire Island National Seashore, 1988).

118. *Strategic Plan for Fire Island National Seashore: Fiscal Year 2001–2005,* pp. 6–9.

119. Ibid.

120. John Rather, "Raw Nerves on Fire Island," *New York Times*, June 23, 2002, Long Island Section, 6–7.

121. Vivian S. Toy, "A Changing of the Guard, and of Policy, on Fire Island," *Newsday*, July 28, 2002.

122. "Fire Island: A Mark Left on Shore," *Newsday*, April 20, 2004, A32.

123. *Land Protection Plan*, 5–6.

Bibliography

"$2,300,000 Erosion Job Gets Underway May 1." *Long Island Press* (Ronkonkoma, NY), January 2, 1957.

"$37 Million Erosion Plan Passes 1st Major Hurdle." *Newsday* (Melville, NY), October 27,1959.

"18,000 Xmas Trees To Fight Erosion." *Newsday* (Melville, NY), January 7,1960.

Babcock, Richard F., and Richard J. Roddewig. *Report to the National Park Service on Fire Island National Seashore.* Chicago: Ross, Hardies, O'Keefe, Babcock & Parsons, 1975.

Baker, Leonard. "Pike Shifts, Backs U.S. Park as Only Hope for Fire Island." *Newsday* (Melville, NY), April 10,1963.

———. "Say Shore Plan Grew From a Moses Acorn." *Newsday* (Melville, NY), June 1963.

Barber, William. "Moses' Pipe Dream? The Master Planner Would Pave Fire Island." *Newsday* (Melville NY), April 4, 1974.

———."Moses' Pipe Dream? The Master Planner Would Pave Fire Island." *Long Island Press* (Ronkonkoma, NY), April 7, 1974.

"Bare $36 Million Plan to Control L.I. Erosion." *Newsday* (Melville, NY), May 9, 1959.

Barron, James. "Fire Islanders Fight U.S. for Their Property." *New York Times*, July 31, 1981.

Bayles, Richard M. *Historical and Descriptive Sketches of Suffolk County and its Towns, Villages, Hamlets, Scenery, Institutions of Important Enterprises with a Historical Outline of Long Island from its First Settlement by Europeans.* Port Jefferson, NY: published by the author, 1874.

Bayles, Thomas R. *Early Years in Brookhaven Town.* Brookhaven, NY: Bicentennial Commission, 1975.

Biderman v. Morton, 497, F.2d 1141 (2d Cir.1974).

Blakelock, Chester R. *History of the Long Island State Parks.* New York: Long Island Forum, 1959.

Bleyer, Bill. "Bringing History to Light." *Newsday* (Melville, NY), April 15, 2003.

———. "A Lifeline for Sailors." *Newsday* (Melville, NY), <http://www.newsday.com/extras/lihistory/5/hs520a.htm> (accessed October 23, 2003).

Blumberg, Barbara. *Celebrating the Immigrant: An Administrative History of the Statue of Liberty National Monument: 1952–1982*. Washington, DC: U.S. Department of the Interior, National Park Service, 1985.

Brevig, Walt. "LIers to Ask What's Score on Erosion." *Newsday* (Melville, NY), August 11, 1959.

Bullington, Neal and David Griese. *Exploring the Sunken Forest, Fire Island National Seashore*. Patchogue, NY: United States Department of the Interior, National Park Service, Fire Island National Seashore, 1977.

Clowes, Ernest S. *The Hurricane of 1938 on Eastern Long Island*. Bridgehampton, NY: Hampton Press, 1939.

Collins, Tom. "State Discloses New Plans for Fire Island Inlet Channel." *Newsday* (Melville, NY), January 11,1960.

Conley, Daniel. "The Short History of Little Pike's Inlet." *Coastlines* 29 (Winter 2000). <http://www.seagrant.sunysb.edu/Pages/Coastlines/Winter00.htm#part2> (accessed January 26, 2003).

Consensus Building Institute. *Conflict Assessment: The Prospects for Building Consensus on Fire Island National Seashore's Vehicle Use Regulations: A Report to the Superintendent of the Fire Island National Seashore*. Cambridge, MA, September 1999.

Daley, William and Clifford Jones. "The Westhampton Dunes Story: A Blueprint for Coastal Management." *Shore and Beach* 68 (January 2000): 2.

DeWan, George. "The 1892 Cholera Panic." *Newsday* (Melville, NY). <http://www.newsday.com/extras/lihistory/histpast/past11.htm> (accessed October 23, 2003).

Dunewood Web Site. <http://www.fairharbor.com/pl_dunewood.htm> (accessed November 3, 2003).

"Erosion Control Project May Begin Within 2 Months." *Long Island Press* (Ronkonkoma, NY), August 16,1963.

"Erosion Job Cost Set at $62 Million." *Newsday* (Melville, NY), October 23, 1957.

Fair Harbor Web Site. <http://www.fairharbor.com> (accessed October 31, 2003).

"Fire Island: A Mark Left on Shore." *Newsday* (Melville, NY), April 20, 2004.

Fire Island Association. "The Fire Island Interim Project Will Not Cause Building on Dunes." White Paper, n.p., April 2000.

———. Response to the Department of Interior's "Development Moratorium," comment on the United States Army Corps of Engineer's Draft Decision Document and Draft Environmental Impact Statement for the Fire Island Interim Plan." Silver Springs, MD: Linowes and Blocher, LLP, March 24, 2000.

"Fire Island Dike Called a Danger to Small Boats." *Long Island Press* (Ronkonkoma, NY), January 14,1960.

Fire Island Guide. Ocean Beach, NY: Trigor Publishing, 1980.

Fire Island National Seashore: Preliminary Report on Findings and Recommendations. Washington, DC: The Conservation Foundation, November 12, 1975.

Fletcher, Larraine A., and Ellen R. Kintz et. al. *Fire Island National Seashore Cultural Resources Inventory*. Boston, MA: Division of Cultural Resources, U.S. Department of Interior, National Park Service, 1979.

Flint, Martha Bockee. *Early Long Island. A Colonial Study*. New York: G. P. Putnam's Sons, 1896.

Focazio, Paul C. and Jay Tanski. "Monitoring Change." *Coastlines* 29 (Winter 2000). <http://www.seagrant.sunysb.edu/Pages/Coastlines/Winter00.htm#part2> (accessed January 26, 2003).

Fowler, Glenn. "The Fire Island Plan Draws Mostly Cheers." *New York Times*, August 1, 1976.

Fresco, Robert. "Fire Island Falls Short of Goals." *Newsday* (Melville, NY), September 15, 1974.

———. "Lifting a Heavy Hand on Fire Island." *Newsday* (Melville, NY), November 13, 1977.

Guidelines for Long Island Coastal Management. Hauppauge, NY: Long Island Regional Planning Board, 1973.

Godfrey, Paul J. "Management Guidelines for Parks on Barrier Beaches." *International Journal of National Parks, Historic Sites and Other Protected Areas*, 2, no. 4 (January–March, 1978): 5–10.

"Governor Avoids Highway Approval." *Island News* (Ocean Beach, NY), July 21, 1962, 1

Grossman, Karl. "Last Long Island Wilderness?" *Long Island Press* (Ronkonkoma, NY), December 5, 1974.

———. "Seashore Plan: Winner in 2d Try." *Long Island Press* (Ronkonkoma, NY), July 28, 1976.

———. "Fire Island Finds Federal Friend," *Long Island Press* (Ronkonkoma, NY), August 23, 1976.

Hamilton, Marsha L. "Shifting Sand: Long Island's Barrier Beaches." *Long Island Historical Review* 16 (Fall 2003/Spring 2004).

Hawes, Elizabeth. "Castles in the Sand." *New York Times Magazine*, July 11, 1993.

———. "Life on the Edge: How Long Can Fire Island Last?" *New York Times Magazine*, July 11, 1993.

"Hearing Scheduled for July 10." *News Review* (Riverhead, NY), June 10, 1962.

Hood, Clifton. "New Studies in Gay and Lesbian History." *Journal of Urban History* (September 24, 1998): 790.

"House Votes 220 Gs for Fire Island Anti-Erosion Project." *Newsday* (Melville, NY), June 3, 1959.

Hurricane Damage Mitigation Plan for the South Shore Nassau and Suffolk Counties, N.Y. Hauppauge, NY: Long Island Regional Planning Board, 1984.

Johnson, Madeleine C. *Fire Island: 1650s–1980s*. Mountainside, NJ: Shoreland Press, 1983.

Kennedy, Shawn G. "U.S. Destroying Squatter Shacks on Fire Island–Dwellers Hope Court Will Delay Park-Site Action." *New York Times*, November 28, 1978.

Koppelman, Lee E. *Report on the Need and Feasibility for County Park Facilities in Suffolk County*. Hauppauge, NY: Suffolk County Planning Department, January 1962.

———. *Comprehensive Open Space Plan*. Hauppauge, NY: Suffolk County Planning Department, May 1964.

———. *Planning for Open Space in Suffolk County*. Comprehensive Plan Series. Volume II. Hauppauge, NY: Suffolk County Planning Department, May 1964.

———. *Land Use, Zoning and Development Control Policies—Fire Island National Seashore*. Hauppauge, NY: Suffolk County Department of Planning, 1975.

———. *Economic Analysis—Fire Island Reach*. Hauppauge, NY: New York Coastal Partnership, 1995.

Krebs, Albin. "Fire Island Seashore—Udall's Latest Plan." *The New York Herald Tribune*, April 9, 1963.

Leatherman, S. P., and J. R. Allen, eds. *Fire Island Inlet to Montauk Point Long Island, New York: Reformulation Study: Executive Summary*. Patchogue, NY: United States Department of the Interior, National Park Service, Fire Island National Seashore, 1985.

LeBoeuf, Randall J. Jr. "The Sunken Forest." *Long Island Forum* (Syosset, NY), April 24, 1971.

Leonard, William Norris and Lee E. Koppelman. *People and Parks*. Hauppauge, NY: Suffolk County Planning Commission, November 1960.

Lipson, Eden. "When the Dancing Had to Stop." *New York Times*, September 5, 1976.

Lonelyville Web Site. <http://www.fairharbor.com/pl_lonelyville.htm> (accessed 31 Oct. 2003).

Mackay, Katurah. "Homes Erode National Seashore." *National Parks Magazine*, July 1999.

Mackintosh, Barry. *The National Park Service: A Brief History*. Washington, DC: National Park Service, 1999 <http://www.cr.nps.gov/history/hisnps/NPSHistory/npshisto.htm> (accessed November 19, 2003).

———. *The National Parks System: Shaping the System*. 5th ed. Washington, DC: National Parks Service, 1985. <http://www.cr.nps.gov/history/online_books/mackintosh1/sts2b.htm> (accessed November 19, 2003).

Maiorana, Ronald. "Park Bill to Pass Fire Island Told." *New York Times,* October 1, 1963.

"Master Plan: A Primer." *Long Island Press* (Ronkonkoma, NY), May 18, 1975.

McHarg, Ian L. *Design with Nature*. Garden City, NY: The Natural History Press, 1969.

Mid-Fire Island Preservation Association newsletter of December 10, 1979.

Moore, Virginia Bennett. "A Possible National Seashore." *National Parks Magazine,* February 1963.

Morris, Tom. "Fire Island: From Pirates to Slavery to Fun in the Sun." *Newsday* (Melville, NY). <http://www.newsday.com/extras/lihistory/spectown/hist0071.htm> (accessed October 23, 2003).

"Moses' Fire Island Plan Okayed By State Shoreline Commission." *Smithtown Star,* August 1, 1962.

"Moses Offers Plan to Prevent New Storm Devastation on the Atlantic Shore." *Long Island Press* (Ronkonkoma, NY), March 26, 1962.

Muller, Robert G. *Long Island's Lighthouses Past and Present*. Patchogue, NY: Long Island Chapter of the US Lighthouse Society, 2004.

Nash, Collin. "Focus on History: William Floyd Estate." *Newsday* (Melville, NY), July 2, 1995.

"New Snag Perils Erosion Control Plans." *Newsday* (Melville, NY), February 26, 1963.

Newton, Esther. *Cherry Grove, Fire Island: Sixty Years in America's First Gay and Lesbian Town*. Boston, MA: Beacon Press, 1993.

Now and for the Future: A Vision for New York's Coast. New York: Recommendations of the Governor's Task Force on Coastal Resources, n.p., November 1991.

Ocean Beach Web Site. <http://www.ocean-beach.com/pl_ob.htm> (accessed October 29, 2003).

"OK Due Today on F.I. Road Plan." *Newsday* (Melville, NY), July 26, 1962.

Omholt, Thore. "Coastal Processes and Shore Protection: Address to the Marine Resources Council Committee on Beach Erosion." *Barrier Beach Dynamics*, n.d.

Panner, Vicky. "Javits, Pike Open Hearing on Fire Island Shore Bill." *L.I. Commercial* Review (Ronkonkoma, NY), October 1, 1963.

Panuzio, F. L. "Atlantic Coast of Long Island." *Proceedings 11th Conference of Coastal Engineers*. Ed. J. W. Johnson. Richmond, CA: Council on Wave Resistance, 1968.

Paquette, Carole. "Fire Island: Accent is on Upgrades, and Just Up." *New York Times*, April 22, 2001.

Pilkey, O. H., O. H. Pilkey, Sr., and R. Turner. *How to Live with an Island*. Raleigh, NC: NC Department of Natural and Economic Resources, 1975.

Point O'Woods Web Site. <http://www.fairharbor.com/pl_pointowoods.htm> (accessed October 31, 2003).

Porterfield, Byron. "Fight to Save L.I. Beaches Will Start Tomorrow." *New York Times,* June 4, 1959.

"Preserving Fire Island." *Newsday* (Melville, NY), September 17, 1974.

Proceedings of the First Long Island Coastal Alliance Conference. N.p., 1990.

"Propose New Inlet at Fire Island." *Long Island Press* (Ronkonkoma, NY), March 15, 1960.

"Proposed Fire Island Inlet May Cause Havoc in Bay." *Long Island Press* (Ronkonkoma, NY), March 21, 1960.

Proposed Long Island South Shore Hazard Management Program. Hauppauge, NY: Long Island Regional Planning Board, 1989.

Public Law 88–587, 88th Congress, S. 1365, September 11,1964.

Public Law 89–244 (All) William Floyd Estate (1965).

Rather, John. "Raw Nerves on Fire Island." *New York Times*, June 23, 2002, Long Island Section.

"Red Tape Not Halting Fire Island Erosion." *Newsday* (Melville, NY), March 5, 1962.

Renner, Tom. "Supervisors Cool to Army Erosion Plan." *Newsday* (Melville, NY), May 12, 1959.

"Representative Stuyvesant Wainwright Called for the Creation of a National Seashore." *Suffolk County News* (Sayville, NY), July 27, 1956.

Revson, James A. "Fire Island: An Island of Extremes." *Newsday* (Melville, NY), July 11, 1985.

Ritch, Carol Lee. "Fire Island Eviction to Proceed." *Newsday* (Melville, NY), November 16, 1978.

"Rocky Says State Has No Funds for Fire Island Road." *L.I. Commercial Review* (Ronkonkoma, NY), February 27, 1963.

Ruhfel, Albert. "The Fire Island Inlet." *Long Island Forum* (Syosset, NY), 34 (1971): 90–93.

Ruzyla, Kenneth. "Effects of Erosion on Barrier-Island Morphology, Fire Island, New York." In *Coastal Morphology*. Ed. Donald R. Coates, 220. Binghamton, NY: State University of New York, 1972.

Saville, T., Jr. "Sand Transfer, Beach Control, and Inlet Improvements, Fire Island Inlet to Jones Beach." *Proceedings of the 7th Conference of Coastal Engineers*. Ed. J. W. Johnson, 785–807. Richmond, CA, 1961.

Schuberth, Christopher. "Barrier Beaches of Eastern America." *Natural History,* June–July 1970:53.

"Seashore Plan Called All Wet." *Long Island Press* (Ronkonkoma, NY), January 20, 1975.

"Seems Fire Island Road's Almost Washed Up! Barbash Happy as Moses Leaves the Long Island State Park Commission." *New York Standard,* January 14, 1963.

"Set Dedication of Forest Preserve." *Newsday* (Melville, NY), August 12,1959.

Shaw, Osborne. Letter to the Editor. *The Long Island Advance* (Patchogue, NY), November 22, 1956.

Soundings: Annual Report, n.p., *2003*:7.

Spencer and Terchunian. "The Sand Thieves of Long Island's South Shore." *Shore and Beach* (July 1997): 8.

Starace, Carl A. "Fire Island Surf Hotel." *Long Island Forum* (Syosset, NY) (August 1970): 156–159.

"State Erosion Unit Details Plan for L.I." *Newsday* (Melville, NY), July 25, 1962.

Stout, August, Jr. *Pictorial Memories of the Moriches Area*. New York: n.p., 1973.

"Stuy Hits Moses Again on Inlet Project." *Newsday* (Melville, NY), April 11, 1960.

"Suffolk Board to Back Fire Island Park." *Newsday* (Melville, NY), April 3,1963.

"Suffolk Stalls Erosion Job Verdict." *Newsday* (Melville, NY), August 6, 1963.

"Suffolk Votes to Buy Tract on Fire Island." *Long Island Press* (Ronkonkoma, NY), August 13, 1963.

"Supervisors Back Fire Island Park, Delay Erosion Plan." *Long Island Press* (Ronkonkoma, NY), April 9,1963.

Maurice Swift, "U.S. to Acquire Land, 31 Homes on Fire Isle." *Newsday* (Melville, NY), n.d.

Taney, N. E. *Geomorphology of the South Shore of Long Island, New York*. U.S. Army Coastal Eng. Research Center, Tech. Memo 128:50.

Tanski, Jay and Henry Bokuniewicz, *Special Study No. 144*. Stony Brook, NY: New York Sea Grant Institute, 1992.

"The Moses Plan." *Long Island* Press (Ronkonkoma, NY), March 25, 1962.

The Regional Plan Association of New York. *The Race for Open Space*. New York: n.p., 1960.

Thompson, Betty. *Administrative History of Fire Island National Seashore.* Patchogue, NY: United States Department of Interior, National Park Service, Fire Island National Seashore, 1977.

Tooker, William Wallace. *The Indian Place-Names on Long Island and Island Adjacent with Their Probable Significations.* New York: G. P. Putnam's The Knickerbocker Press, 1911.

"Top Army Engineer Calls Fire Island Project a Success." *Long Island Press* (Ronkonkoma, NY), May 12, 1960.

Toy, Vivian S. "A Changing of the Guard, and of Policy, on Fire Island." *Newsday* (Melville, NY), July 28, 2002

Tufton, Mason J. "Early History of Saltaire." In Saltaire Citizen's Advisory Association, 20; Saltaire Web Site. <http://www.fairharbor.com/sa/pl_saltaire.htm> (accessed October 31, 2003).

Tuppin, Andrea. "Staying On." *New York Times*, February 14, 1999, Long Island Sunday Supplement.

"Udall Reaffirms Support for Fire Island Seashore." *Long Island Commercial Review* (Ronkonkoma, NY), April 2, 1963.

United States of America against certain lands located in the County of Suffolk, State of New York, et al., 67c705; 68c838.

United States of America, Plaintiff—Appellee against Certain Lands Located in the County of Suffolk, State of New York, et. al., Defendants—Appellants Brief for Appellee, on behalf of Edward R. Korman, United States Attorney for the Eastern District of New York.

United States. Army Corps of Engineers New York District. *Final Environmental Impact Statement for Fire Island Inlet to Montauk Point.* New York, September 1977.

———. "Fire Island Inlet to Montauk: Fact Sheet." New York: n.p., 2003.

———. "Fire Island to Montauk Point Reformulation Study: Vision Statement." Power Point Presentation, 2003.

———. *Fire Island to Montauk Point, New York, Combined Beach Erosion Control and Hurricane Protection Project.* 1959.

United States. Department of Interior comments to United States Army Corps of Engineers. *Draft Environmental Impact Statement.* February 29, 2000.

United States. Department of Interior, National Parks Service, Denver Service Center, Denver. *Conceptual Master Plan For Fire Island National Seashore*, 1973.

United States. Department of the Interior, National Park Service. *Our Vanishing Shoreline—Seashore Recreation Area Survey of Atlantic and Gulf Coasts.* 1963.

———. *Fire Island National Seashore: A Proposal.* Washington, DC: n.p., 1964.

———. *Fire Island National Seashore Draft Statement for Management.* n.p., 1979.

———. *Fire Island National Seashore Statement for Management.* Patchogue, NY: n.p., 1979.

———. *National Park Service Moves Forward With Management of Personal Watercraft Use.* Washington, DC: Government Printing Office, April 16, 2002

United States. Department of the Interior, National Park Service, Fire Island National Seashore. *Deer and People at Fire Island: How You Can Help Protect the Health of Deer and People*. Patchogue, NY, n.d. <http://www.nps.gov/fiis/deerpeople/deer.html>.

———. *The Master Plan: Fire Island National Seashore*, May 1965.

———. *A Study of Visitor Use Satisfactions*. Patchogue, NY: n.p., October 1969.

———. *Draft Environmental Impact Statement, Master Plan for Fire Island*. Patchogue, NY: n.p., 1975.

———. *Draft Master Plan for Fire Island*. Patchogue, NY: n.p., 1975.

———. *Draft General Management Plan Summary*. June 1976.

———. *Final Transportation Study*. Patchogue, NY: n.p., 1977.

———. *Description of Selected Alternatives Site Development and Facilities: William Floyd Estate, Fire Island National Seashore*. Patchogue, NY: n.p.,1979.

———. *Land Protection Plan*. Patchogue, NY: n.p., 1984.

———. *Annual Narrative Report of Superintendents: Fire Island National Seashore*. Patchogue, NY: n.p., 1987.

———. *Organization Review*. Patchogue, NY: n.p., 1988.

———. *Annual Narrative Report of Superintendents: Fire Island National Seashore*. Patchogue, NY: n.p., 1994.

———. *Environmental Assessment: Amendment to Wilderness Area Management Plan (1984)*, Patchogue, NY: n.p., 1996.

———. *Environmental Assessment for Endangered Species Habitat Management*. Patchogue, NY: n.p., 1998.

———. *Fire Island National Seashore Waterborne Transportation System Plan*. Patchogue, NY: n.p., February 2001.

———. *Negotiated Rulemaking Advisory Committee for Off-Road Driving Regulations at Fire Island National Seashore: Final Consensus Agreement*. Patchogue, NY: n.p., August 22, 2003.

———. *Strategic Plan for Fire Island National Seashore: 2001–2005*. <http://www.nps.gov/fiis/stratplanFY01-05.htm> (accessed May 5, 2004).

———. *Mosquito Surveillance and Management Protocol: Fire Island National Seashore—2002* and *Mosquito Action Plan*, 2002, < http://www.nps.gov/fiis/MSMPfy02.htm> (accessed May 5, 2004).

United States. Department of the Interior, National Parks Service, Northeast Regional Office and Eastern Office of Design and Construction. *Fire Island National Seashore: A Proposal*. n.p. September 1963.

United States. Department of Interior, National Park Service, U.S. Bureau of Outdoor Recreation. *Fire Island National Seashore: A Proposal*. March 1964.

U.S. Congress. *Congressional Record*. 88th Cong., 1st sess., Feb. 11, 1963. HR 3693. *Fire Island National Seashore Act of 1963*.

U.S. Congress. *Congressional Record*. 88th Cong., 1st sess., Mar. 19, 1963. H.R. 4999. Proposed by Ryan (D) NY to provide for the establishment of the Fire Island National Seashore.

U.S. Congress. *Congressional Record*. 88th Cong., 1st sess., April 25, 1963. S. 1365. Proposed by Javits (R) NY to establish the Fire Island National Seashore.

U.S. Congress. *Congressional Record.* 88th Cong., 1st sess., May 7, 1963. H.R. 6111. Proposed by Grover (R) NY to establish the Fire Island National Seashore.

U.S. Congress. *Congressional Record.* 88th Cong., 1st sess., May 9, 1963. H.R. 6213. Proposed by Wydler (R) NY to establish the Fire Island National Seashore.

U.S. Congress. *Congressional Record.* 88th Cong., 1st sess., June 11, 1963. H.R. 6934. Proposed by O'Brien (D) NY to provide for the establishment of the Fire Island National Seashore.

U.S. Congress. *Congressional Record.* 88th Cong., 1st sess., June 11, 1963. H.R. 6936. Proposed by Ryan (D) NY to provide for the establishment of the Fire Island National Seashore.

U.S. Congress. *Congressional Record.* 88th Cong., 1st sess., June 18, 1963. H.R. 7107. Proposed by Pike (D) NY to provide for the establishment of the Fire Island National Seashore.

U.S. Congress. *Congressional Record.* 88th Cong., 1st sess., June 26, 1963. H.R. 7297. Proposed by Kelly (D) NY to provide for the establishment of the Fire Island National Seashore.

U.S. Congress. *Congressional Record.* 88th Cong., 1st sess., July 1, 1963. H.R. 7359. Proposed by Keogh (D) NY to provide for the establishment of the Fire Island National Seashore, in New York.

U.S. Congress. *Congressional Record.* 88th Cong., 1st sess., Oct. 9, 1963. H.R. 8035. Proposed by Pike (D) NY to authorize the acceptance of the donation of the William Floyd Estate in Suffolk County, New York for addition to the Fire Island National Seashore.

U.S. Congress. *Congressional Record.* 96th Cong., 2nd sess., August 19, 1980. S. 304. Proposed by Senators Javits (R) NY and Moynihan (D) NY to designate the "Otis Pike-Fire Island High Dune Wilderness" on August 19, 1980 (companion H.R. 7814 introduced by Representatives William Carney and Norman Lent).

Vetter, John F., and Bert Salwen, *Report of an Archeological Reconnaisance of Fire Island, Suffolk County, New York*. Patchogue, NY: U.S. Department of Interior, National Park Service, Fire Island National Seashore, 1979.

Vollmer Associates. *Fire Island National Seashore: Preliminary Transportation Study, Work Directive No. 4–0070–74–4,* July 26, 1974.

Weeks, George L. Jr. *Isle of Shells*. New York: Buys Bros., 1965.

Whitehouse, Beth. "Fire I. Fights Over Coveted Permits." *Newsday* (Melville, NY), March 6, 1994.

———. "Shack to Chic on Fire Island." *Newsday* (Melville, NY), July 7, 1993.

Wittlock, Lavern A., Sr. "Conjecture on Origin of Fire Island." *Long Island Forum* (Syosset, NY), September 1973, 165.

Yasso, Warren E. and Elliott M. Hartman, Jr. "Beach Forms and Coastal Processes." *MESA New York Bight Atlas Monograph 11*, 21. New York: New York Sea Grant Institute, 1975.

Index